THR  BUS

'Thoughtful, provocative and essential reading for our economic moment'
Joi Ito, director, MIT Media Lab

'We've optimized for growth. But have we lost our way? As an economy? As a community? As a society with a value proposition that doesn't make sense on a human or economic level? Rushkoff asks questions that matter. A challenging and necessary read'
Sherry Turkle, author of *Reclaiming Conversation*

'Every great advance begins when someone sees that what everyone else takes for granted may not actually be true. Douglas Rushkoff questions the deepest assumptions of the modern economy and blazes a path toward a more human-centred world'
Tim O'Reilly, founder, O'Reilly Media

'Douglas Rushkoff is a true digital visionary. Read this rousing call to reboot our society from the bottom up before it's too late'
Astra Taylor, film-maker and author of *The People's Platform*

'In what could be seen as a crisis, Rushkoff shares his smart, optimistic and pragmatic perspective about how both businesses and consumers can reimagine today's current economic operating system in the digital age – and prosper'
Bonin Bough, chief media and e-commerce officer, Mondelēz

ABOUT THE AUTHOR

Douglas Rushkoff is the bestselling author of *Present Shock*, as well as a dozen other books on media, technology and culture, including *Program or Be Programmed* and *Life Inc.* Named one of the world's ten most influential thinkers by MIT, he has made documentaries for PBS *Frontline*, including *Generation Like* and *The Merchants of Cool*, and he is a professor of media theory and digital economics at Queens College, CUNY. He lives in New York and lectures about media, society and economics around the world.

THROWING ROCKS AT THE GOOGLE BUS

How Growth Became the Enemy of Prosperity

Douglas Rushkoff

PORTFOLIO
PENGUIN

PORTFOLIO PENGUIN

UK | USA | Canada | Ireland | Australia
India | New Zealand | South Africa

Portfolio Penguin is part of the Penguin Random House group of companies whose addresses can be found at global.penguinrandomhouse.com.

First published in the United States of America by Portfolio/Penguin, a member of Penguin Group (USA) Inc. 2016
First published in Great Britain by Portfolio Penguin 2016
001

Printed in Great Britain by Clays Ltd, St Ives plc

A CIP catalogue record for this book is available from the British Library

ISBN: 978–0–241–00441–8

Penguin Random House is committed to a sustainable future for our business, our readers and our planet. This book is made from Forest Stewardship Council® certified paper.

For Barbara and Mamie

Contents

Introduction

WHAT'S WRONG WITH THIS PICTURE?

One December morning in 2013, residents of San Francisco's Mission District laid their bodies in front of a vehicle to prevent its passage. Although acts of public protest are not unusual in California, this one had an unlikely target: the Google buses used to ferry employees from their homes in the city to the company's campus in Mountain View, thirty miles away.

As the photos and live updates from the scene filled my social media feeds, I wasn't sure how to react. Google was, in most ways, the ultimate Internet success story—a college dorm-room experiment that had blossomed into one of the world's most powerful technology giants and created thousands of jobs—all while striving to do no evil. The company's stellar growth revived more than a few economic sectors, as well as a few neighborhoods. And for a while, everybody was happy with the way things were going. We all got free search and e-mail. Bloggers got paid to put ads on their Web sites, kids shared in the revenue from YouTube videos, and the Mission District got a bit trendier and safer as hipsters and tech professionals moved in, new coffee and book shops opened, lofts were constructed, and property values went up. Growth is good—at least for those doing the growing.

But the influx of Googlers to San Francisco's most historic neighborhoods also raised rents, forcing out longtime residents and small businesses that were not participating in all this growth. Google's air-conditioned tour buses epitomized the seeming invasion—like space transports taking aliens to and from their mother ship. Adding insult to injury, Google was now using publicly funded bus stops as loading stations for its very private transportation system. Rents close to those bus stops were 20 percent higher than those in comparable areas,[1] which were themselves doubling every few years to accommodate not only Google's employees but those of Facebook, Twitter, and the other Silicon Valley darlings.

And so on the same day Google's stock happened to be reaching a new high on Wall Street, a dozen scrappy, yellow-vested protesters managed to paralyze one of the tech giant's now infamous buses. Onto its side they plastered an Instagram-friendly banner that read "Gentrification & Eviction Technologies" in a perfect, multicolored Google font. Tapping into growing skepticism over the unequally distributed benefits of the tech boom, the image spread like wildfire. At least some small part of me smiled in solidarity with their critique.

A few weeks later, there was nothing to smile about. Protesters in Oakland were now throwing rocks at Google's buses and broke a window, terrifying employees. Sure, I was as concerned about the company's practices as anyone, and frustrated by the way Silicon Valley's rapid growth seemed to be displacing instead of enriching the people of San Francisco and beyond. But I also had friends on those buses, trying to make a living off their hard-won coding skills. They may have made $100,000 a year, but they were stressed-out, perpetually monitored, and painfully aware of their own perishability. "Sprints"—bursts of round-the-clock coding to meet deadlines—came ever more frequently as new, more ambitious growth targets replaced the last set.

We may all be on the same side here. Google workers are less the beneficiaries of an expanding company than they are its rapidly consumed resources. The average employee leaves within a year[2]—some to accept better positions at other companies but most of them simply to break free of the constant pressure to perform. Taking the bus gives them more time to work or just relax instead of driving. They are human beings.

For its part, Google is relieving the freeways and the environment of thousands of commuter cars. Unlike many other companies in the Bay Area, which give only lip service to environmentalism or, at best, organize car pools, Google offers a shuttle program that saves more than 29,000 metric tons of CO_2 per year. Since when has doing the right thing become the wrong thing?

There is something troubling about the way Google is impacting the world, but neither its buses nor the people in them are the core problem; they're just the easy target. Google's employees are not oblivious to the increasing poverty outside the bus windows on their way to work. If anything, such sights only make these workers cling to their jobs all the more desperately, leaving them less likely to question the deeper processes at play. They do want to become millionaires—but not so that they can live a life of luxury. In a country without a strong social safety net, workers are told that they have to become millionaires or else face penury as soon as they retire or, worse, get sick. A typical online retirement calculator insists that a person who earns $50,000 a year will require at least $1.5 million to retire at age sixty-seven, and a single unexpected medical bill can turn any of us into one of America's 1.7 million annual health-related bankruptcies.

Not even Google's investors, officers, or the infamous 1 percent are to blame for the growing inequalities of the digital economy. Silicon Valley executives and venture capitalists are simply practicing capitalism as they learned it in business school and, for the most part, meeting their legal obligation to the shareholders of their companies. Sure, they are getting wealthier as the rest of us struggle, and yes, there's collateral damage associated with the runaway growth of their companies and stocks. But they are as stuck in this predicament as anyone; many CEOs understand that meeting short-term growth targets is not in the best long-term interests of their companies or their customers, but they are themselves caught up in a winner-takes-all race for dominance against all the other digital behemoths. It's grow or die. So each tech company must become as intrusive, extractive, divisive, time-consuming, wasteful, expensive, job killing, exploitative, and manipulative as the next one. As for their impatient shareholders, well, they are the likes of us: we are the ones holding these very stocks in our own 401(k) and college savings plans, counting on them to go up, and selling them if they don't. None of this worked out as we thought it would, and we're all frustrated by the results.

But there's no easy place to draw the battle lines or enemy at whom to hurl the rocks. That's because the conflict here is not really between San Francisco residents and Google employees or the 99 percent and the 1 percent. It's not even stressed-out employees against the companies they work for or the unemployed against Wall Street so much as everyone—humanity itself—against a program that promotes *growth* above all else.

We are caught in a growth trap. This is the problem with no name or face, the frustration so many feel. It is the logic driving the jobless recovery, the low-wage gig economy, the ruthlessness of Uber, and the privacy invasions of Facebook. It is the mechanism that undermines both businesses and investors, forcing them to compete against players with digitally inflated poker chips. It's the pressure rendering CEOs powerless to prioritize the sustainability of their enterprises over the interests of impatient shareholders. It is the unidentified culprit behind the news headlines of economic crises from the Greek default to skyrocketing student debt. It is the force exacerbating wealth disparity, increasing the pay gap between employees and executives, and generating the power-law dynamics separating winners from losers. It is the black box extracting value from the stock market before human traders know what has happened, and the mindless momentum expanding the tech bubble to proportions dangerously too big to burst.

To use the metaphor of our era, we are running an extractive, growth-driven economic operating system that has reached the limits of its ability to serve anyone, rich or poor, human or corporate. Moreover, we're running it on supercomputers and digital networks that accelerate and amplify all its effects. Growth is the single, uncontested, core command of the digital economy.

Classical economists and business experts have been of little help. This is because they tend to accept the growth-based economy as a preexisting condition of nature. It is not. The rules of our economy were invented by particular human beings, at particular moments in history, with particular goals and agendas. By refusing to acknowledge the existence of this man-made landscape and our complicity in perpetuating it, we render ourselves incapable of getting beneath its surface. We end up transacting and living at the mercy of a system.

We must instead take a good look at the underlying assumptions of the marketplace we're busy digitizing and ask ourselves if they are still relevant to our situation before we let our computers and networks run with them. Perhaps ironically, only by thinking like programmers can we adapt the economy to serve human beings instead of the quite arbitrary but deeply embedded ideal of growth.

Until we do, we remain incapable of properly seeing, much less changing, the functioning of our businesses or the economy in which they operate. We are destined to repeat the same old mistakes. Only this time, thanks to the speed and scale on which digital business operates, our errors threaten to derail not only the innovative capacity of our industries but also the sustainability of our entire society. People throwing rocks at the Google bus will be remembered as the tremor before the quake.

Or we may come to our senses and choose a different path. We are at a critical crossroads. Every businessperson, employee, entrepreneur, or creator reading this book understands that we are all operating on borrowed time and borrowed money. We need to make a choice. We can continue to run this growth-driven, extractive, self-defeating program until one corporation is left standing and the impoverished revolt. Or we can seize the opportunity to reprogram our economy—and our businesses—from the inside out.

In doing so, we can begin to benefit from new, more distributed modes of value creation and exchange—the sorts of wealth generation we see on trading sites, on peer-to-peer lending networks, on user-owned platforms, or even in the games and apps programmed by college kids on laptops and then brought directly to market.

That's the economy most of us early net users envisioned back in the 1980s when we first got our hands on these new tools. But by the early 1990s, this human-centered vision of a networked marketplace was replaced by another vision of digital business—the one espoused by the libertarian founders and early writers of *Wired* magazine and the corporate-sponsored futurists of Cambridge, Massachusetts, many of whom were the very same people. They looked at digital technology and saw in it a way to revive the imperiled securities markets as well as to restore faith in the notion of an infinitely expanding economy.

After the biotech crash of 1987, many feared that the half century of unparalleled growth following World War II might finally be over. But now digital technology was to return the NASDAQ to its former glory and beyond. Indeed, just when it looked like we had reached the limits of the physical world to supply us with more opportunities for growth, we discovered a virtual world from which to extract still more value. According to the new pundits, this new digital economy would augur a "long boom"[3] of economic growth: a digitally amplified, speculative economy that could literally expand forever.

We optimized our platforms not for people or even value but for growth. So instead of getting more free time, we ended up getting less. Instead of getting more varieties of human expression and interaction, we pushed for more market-friendly predictability and automation. Technologies were prized most for their ability to extract value from people in terms of "eyeball hours" and the data that could be derived from them. As a result, we have ended up in an always-on digital landscape, constantly pinged by updates and enduring a state of perpetual emergency interruption—what I call "present shock"—previously known only to 911 operators and air traffic controllers.

We are developing our new technologies not for the betterment of humanity or even our businesses but to maximize the growth of the speculative marketplace. And it turns out that these are not the same thing.

On the bright side, millionaires and billionaires are being minted. Maybe not enough to compensate for the millions of people displaced from their jobs or disconnected from prosperity by the same mechanisms, but they are inspiring all the same. They come in on every new tech wave, whether it's Web sites, blogging, search, social media, file exchange, photo sharing, messaging, or cloud services . . . each new Internet trend produces a new billionaire for the front page of the *Wall Street Journal*. They are a different sort of billionaire: "paper" billionaires, whose worth is measured in stock, not profits. That's because, for the most part, the Internet companies they run don't really have profits—certainly not ones commensurate with their market capitalizations. They become success stories only on the day the founders and investors make their exit—that is, sell their stake for something more real, like another company or just plain money.

That's the game most of us now call the digital economy. It is accepted unquestioningly, because it stokes the flames of growth—however artificially. Companies with new technologies are free to disrupt almost any industry they choose—journalism, television, music, manufacturing—so long as they don't disrupt the financial operating system churning beneath it all. Hell, most of the founders of these digital companies don't seem to realize this operating system even exists. They are happy to challenge one "vertical" or another, but the last thing they do when they've got a winner is challenge the rules of investment banking, their own astronomical valuation, or the IPO through which they cash out. Winning the digital growth game is less a new sort of prosperity than it is a new way to execute business as usual: old wine in a new bottle. It's not that making money is so wrong; it's that the premises of venture capital and the stock market—as well as their real effects—are never even questioned. The winners have, in some fundamental way, been duped.

That's why when I saw Twitter cofounder Evan Williams on the front page of the *Wall Street Journal* in a photo taken the morning of his IPO, I felt happy for him yet a bit saddened. Just under his chin they had printed the number $4.3 billion—the amount of money he made that day—making him by far the richest person I've ever known personally.[4] This was the kid who started Blogger, struggled to keep it afloat, and then made his very first millions by selling it to Google. Here he was now—like the boy at the county fair getting a ribbon for growing an inconceivably gargantuan pumpkin—one of the wealthiest men in the world. But at what cost?

Evan had disrupted journalism with the blog, and newsgathering with the tweet, but now he was surrendering all that disruption to the biggest, baddest industry of them all. When you're on the front page of the *Wall Street Journal*, receiving applause from all those guys in suits, it's not usually because you've done something revolutionary; it's because you have helped confirm financial capital's centrality to the whole scheme of human affairs. As the dealer at the casino shouts loudly enough for everyone to hear, "We have a winner!" The growth game is still working, so place your bets.

Evan and his partners successfully turned Twitter into a publicly traded, multibillion-dollar company and in the process sacrificed a potentially world-changing app to the singular pursuit of growth. Here was arguably the most powerful social media tool yet developed—from organizing activists in the Arab Spring and Occupy Wall Street movements to providing a global platform for citizen journalists and presidential candidates alike. And it wasn't particularly expensive to create or maintain. It certainly didn't require a multibillion-dollar cash infusion in order to keep functioning.

Having taken in this much new capital, however, Twitter now needs to produce. It must *grow*. As of this writing, the $43 million Twitter *profited* last quarter is considered an abject *failure* by Wall Street. In 2015, Twitter investors complained* that the company was too far from reaching its *"100x" growth potential* and forced out the CEO. Shareholders are demanding that Twitter find better ways of monetizing its users' tweets, whether

---

* Most notably Chris Sacca, in public statements.

by injecting advertisements into people's feeds, mining their data for marketing intelligence, or otherwise degrading the utility of the app or the integrity of its community. Whatever actually may have been disruptive about Twitter will now have to be made less so.

It's not that Twitter isn't successful; it's just not successful enough to justify all the money investors have pumped into it. There was already enough revenue for the employees to be happy, the users to be served, and even the original investors to be well compensated in an ongoing way. But there may never be enough to satisfy shareholders who expect to win back one hundred times their initial $20 billion bet. To do that, Twitter must grow into a corporation bigger than the economy of many entire nations. Isn't that a bit much to ask of an app that sends out messages of 140 characters or less?

This disproportionate relationship between capital and value—or invested money versus actual revenue—is the hallmark of the dominant digital economy. If anything, the digital economy has laid bare the process by which cash, labor, and productive assets from the real, transactional marketplace are extracted and converted into frozen capital—all in the name of growth. Once money has been "captured" in a stock price, it tends to just sit there as if in a bank vault. This, in turn, puts pressure on the company to make more money, faster, in order to justify the new total value of all the stock. The disparity between a company's net worth and its revenues gets even more extreme. Strangely enough, the companies do keep growing, but they don't create or produce any value.

This part isn't new. In fact, for the past seventy years or more, according to economists at the Deloitte Center for the Edge,[5] corporate profits over net worth have been steadily *decreasing*. Companies are accumulating money and assets faster than they can utilize them. This doesn't mean that corporations aren't still rich. It just means they don't know how to apply the assets they already have. They have grown too much for their own—or anybody else's—good.

What is new here is that by applying our technological innovations to growth above all else, we have set in motion a powerfully destabilizing form of digitally accelerated capitalism. We are worsening the disparity between rich and poor, punishing those who actually work for a living, losing human control over the capital markets, and letting shortsighted investors stifle long-term innovation. For many, digital technology is just an invitation to game the markets in new ways, creating increasingly abstract and ultrafast instruments for beating the system instead of creating value.

There are better approaches to achieving prosperity in a digital business landscape. If we can get over our addiction to growth, we have the potential to move toward a much more functional, even compassionate economic system that favors money flow over accumulation and rewards people for creating value instead of simply extracting it. But first we have to acknowledge and address our current state of affairs, and the instability and insecurity under which so many of us are trying to function.

For many of us, the current system, however convoluted, is better than nothing, and changing to one in which we must create real value is frightening. Most people are not cultural creatives capable of launching a business on Etsy, programming a new iPhone app, or growing artisanal organic yams. We work in cubicles managing spreadsheets, calculating sales targets, and budgeting ad spends—or in retail stores, on factory floors, and in warehouses—doing jobs that may have no application or value outside that single corporate setting. We are simply fighting to stay employed, pay our mortgages, save for our kids' college, and make sure we have something left for retirement. And in spite of the digital boom—or maybe because of it—it's getting harder to do any of those things. The path to a better tomorrow must first make today a bit easier on us all. We have to take practical, baby steps.

But to begin, we must accept the story of an infinitely expanding market for the myth that it is. As we'll see, growth may be a requirement of interest-bearing currency and venture capital, but it's not a requirement of business or commerce. If we are going to do something better than digitize the industrial economy and amplify the worst of its effects, we have to recognize how the expansionist agenda of our colonial forebears operates in today's more limited environment, and why it is at cross-purposes with our potential goals as a digitally enabled society.

Neither individuals, small businesses, corporations, nor even whole governments need to live and die by their rate of growth. This is not bad news but good news. And the sooner we accept this, the sooner we'll all be off the hook. Then, and only then, will we be capable of ushering in the sort of economy we deserve. Ongoing, sustainable, and distributed prosperity is simpler than it sounds, and well within our reach. It could be our new normal.

This is the true promise of a digital economy. Uncovering the unacknowledged operating system driving our businesses is just the first step. Mustering the willingness to do business differently or even change that system instead of continuing to feed its growth is next. We won't do this all at once. We can't just install an update, as we do on a smartphone, however much we might like to believe in technological fixes for the world's problems. There's no algorithm for this. There's only slow, incremental change, enacted consciously but differently by all sorts of people and institutions. That's the difference between an industrial society and a digital one: one-size-fits-all solutions that take over the entire planet no longer work. Instead, a broad set of distributed solutions coexist, in many places and on many scales at once.

But they all come down to rejecting the notion that the only healthy career, company, or economy is one that grows at a rate defined arbitrarily by a bank, a group of investors, or the startup ethos now so dominant in Internet culture. As members of a digital society, we are uniquely positioned to strive toward a more sustainable, steady state of distributed wealth.

Here's how.

Chapter One

REMOVING HUMANS FROM THE EQUATION

DIGITAL INDUSTRIALISM

Back in the late 1980s, before I could make a living as an author, I used to type for my rent money. A bunch of us aspiring writers would go into a Park Avenue law firm after it closed at night, put on earphones, and transcribe hours of depositions—mostly from factory workers describing accidents they had on the job, supervisors explaining what the worker had done wrong, or doctors detailing why a finger or toe had to be amputated. It was gruesome stuff, the human collateral damage of industrial production.

We "word processors" got paid more per hour than most of the workers whose cases we mindlessly chronicled. That's because we were computer literate at a time when almost no one knew how these machines worked. As writers, we knew the ins and outs of then-obscure software such as WordPerfect and could pull in thirty bucks an hour (a king's ransom in the slacker era) helping "real" businesses transition to the digital age.

On my way into the typing pool each evening, I would pass by an artsy young phone receptionist from the day shift packing up to leave. We'd exchange pleasantries and obscure music and comics references as I worked up the courage to ask her out. But before I even had the chance, she was replaced by a computer. What digital giveth, digital taketh away.

Like many other companies in the late 1980s, the law firm had adopted a new technology known as the auto attendant. From then on, instead of being greeted and routed by a live human being, callers would get the now-familiar computer-operator script: "If you know your party's extension . . ." It was probably most people's first real experience interacting with a computer. But the whole innovation always struck me as a step backward—and not just because it cost me my chance with the artsy receptionist.

From the company's side, it's simply a cost-saving measure. After an initial technological investment, they can dispense with the operators, as well as the paychecks, health benefits, air-conditioning, sexual harassment suits, and other costs associated with human employees. Reducing the bottom line is a great way to create the illusion of top-line growth.

It's the callers who pay the price. Most of us are all too familiar with the frustration of winding down a series of menus, entering in all our numbers and dates and codes, only to get stuck in an endless loop or, worse, disconnected. Any efficiency gained by the company is more than offset by the efficiency lost to everyone else. Sure, a company's shareholders may be happy about the reduction in employees—right until they're trying to reach someone in investor relations on the phone and find themselves as fed up as everyone else.

In essence, the cost of the single company's receptionists has been externalized to all the callers, costing more total time and money to the entire network of businesses and customers. And since every company is doing it, refusal to go digital amounts to a competitive disadvantage. Resistance to this automation is so novel, in fact, that some banks and cable-TV providers now advertise the fact that they have *real humans* answering the phone.

But in the longer term, and especially in a highly connected digital economy, nothing is external. The time and expense that a company passes on to its customers, suppliers, and vendors eventually come right back in the form of reduced sales volume, higher costs, and less feedback, respectively. A company should want its human customers to *enjoy* interacting with it. We should want to cost our suppliers as little as possible so that they can in turn offer us the best prices. And we should want to interact directly with our vendors, so we know what exactly is happening on the front lines. A living human interface provides better intelligence than any survey a company can offer after the fact. Automated menu items dump every incoming human request or problem into a preconfigured category, like bins on an assembly line. The call that requires an unforeseen category, well, that's the customer a company loses, and the intelligence that slips through the cracks of a hopelessly industrial-age approach to digital technology in the workplace.

The headlines, just like the displaced workers themselves, blame computers for this conundrum. But it's not the fault of digital technology at all; rather, it's the fault of a digitally charged business model that stresses the efficiency and growth of companies at the expense of the human beings they should be serving. And serving humans isn't merely some New Age value system that needs protecting against the realities of real-world business. It is the fundamental reason for business in the first place: to create ongoing value for customers, employees, and owners.

Somehow, growth has become an end in itself—the engine of the economy—and human beings have come to be understood as impediments to its functioning. If only people and our idiosyncratic demands could be eliminated, business would be free to reduce costs, increase consumption, extract more value, and grow bigger. This is one of the primary legacies of the industrial age, when the miraculous efficiency of machines appeared to offer us a path to infinite growth—at least to the extent that human interference could be minimized. Applying this same ethos in a digital age means replacing the receptionist with a computer, the factory worker with a robot, and the manager with an algorithm. It's all just a new, digital way of running the same old program.

Counterintuitively, perhaps, the greater opportunity of our new digital environment is to retrieve the human values we left behind in that last big technological upheaval. The word *digital* itself refers to the digits—the ten fingers we humans use to build, to count, and to program computers in the first place. That we should now be witnessing a renaissance in makers, crafts, and artisanal production is no coincidence. The digital landscape encourages production from the periphery, lateral trade, and the distribution of wealth. Instead of depending on centralized institutions for sustenance, we begin to depend on one another.

The transparency offered by the digital media landscape has the potential to lay bare the workings of industrialism. Meanwhile, digital technology itself provides us the means to reprogram many business sectors from the ground up, and in ways that distribute value to their many human stakeholders instead of merely extracting it. But doing so requires a rather radical reversal in the way we evaluate business processes and the purpose of technology itself. By reducing human beings to mere cogs in a machine, we created the conditions to worship growth over all other economic virtues. We must reckon with how and why we did this.

MASS MASS MASS

For a happy couple of centuries before industrialism and the modern era, the business landscape looked something like Burning Man, the famous desert festival for digital artisans. The military campaigns of the Crusades had opened new trade routes throughout Europe and beyond. Soldiers were returning from faraway places after having been exposed to all sorts of new crafts and techniques for building and farming. They even copied a market they had observed in the Middle East—the bazaar—where people could exchange not only their goods but also their ideas, leading to innovations in milling, fabrication, and finance.

The bazaar was a peer-to-peer economy, something along the lines of eBay or Etsy, where attention to human relationships and reputations promoted better business. There was no middleman, no central platform through which exchanges were conducted, except for the appointed time and place of the bazaar itself.[1] Since people transacted back and forth, all sorts of interdependencies developed that in turn fostered more and better commerce. Pete the wainwright bought oats from Joe the oat seller, who needed to provide a good product not simply because he wanted to keep a customer but because he needed good wagon wheels. To give Pete bad oats meant risking more than future business; it meant that the craftsperson making Joe's wife's wagon wheels would be sick on the job. This was a bound community of commerce, where transactions were informed by a multiplicity of values.

The quality of goods and services was maintained by a system of guilds covering each of the major trades. It wasn't a perfect scheme, as guilds often favored the children of existing members, but it was characterized less by competition among members than by the standardization of prices, the training of apprentices, and the exchange of best practices. Members of a guild could decide to make a rule, say, to take Sundays off. This would ensure a shorter workweek for all members without putting anyone at a competitive disadvantage.

Thanks to the emergence of the bazaar, Europe in the late Middle Ages enjoyed one of the most rapid economic expansions in history. For the first time in many centuries, the economy *grew.* People ate more, worked less, and became quite healthy—and not just by the standards of that era.[2] The problem was that while the merchant class was gaining wealth, the aristocracy was losing it. Noble families had enjoyed the spoils of feudalism for centuries by passively extracting the value of peasants who worked the land. They never worried about growth because they didn't need to. Things had always been just fine with them as the lords over everyone.

As the new trading economy grew, however, all this began to change. With many former peasants going into business for themselves, the aristocracy lost its monopoly over value creation. The people's economy was growing while the aristocracy's remained stagnant or even shrank. The nobles had no way to keep up. They looked at this new phenomenon of growth and wanted some of it for themselves. They got their growth, but through forced and artificial means. Where the growth of the peasant economy could be considered natural, or even appropriate, the aristocracy's efforts to usurp it were less so. It's one thing for growth to help peasants achieve subsistence. It's another for those who already own pretty much everything to use growth purely as a means to prevent others' enrichment. But that's exactly what happened.

The nobles still had the power to write the law, and in a series of moves that took place in different countries at different times, they taxed the bazaar, broke up the guilds, outlawed local currencies, and bestowed monopoly charters on their favorite merchants. In exchange for stock, kings granted certain companies exclusive control over their industries. The peer-to-peer, or "P2P," nature of the economy changed—not overnight, but over a couple of centuries—to the top-down economy we know today.[3]

Instead of making and trading, craftspeople had to seek employment from one of the chartered monopolies. Instead of selling their wares, people now sold their hours. Counterintuitively, perhaps, business owners learned to seek out the least qualified workers. A skilled shoemaker might demand pay befitting his expertise. An immigrant seeking day labor could be gotten on the cheap and was easily replaced by another if he protested his hours, conditions, or compensation. But how could a bunch of unskilled workers create a viable product? Welcome to the industrial age.

What we now call industrialization was actually an extension of the aristocracy's effort to usurp the growth it witnessed in the peasants' marketplace and to imitate it by other means. Industry was really just the development of manufacturing processes that required less skill from human laborers. Instead of having to learn how to make shoes, each worker could be trained in minutes to do one tiny part of the job. In the long run, many industrial processes have ended up more efficient than production by individual craftspeople, but that's most often because their total costs are hidden or externalized to others. (The government pays for wars to procure cheap oil and roads to convey mass-produced products, while we all pay for the environmental stresses caused by corporate agriculture, and so on.) Prices may be low, but the *costs* are high. It was never really about efficiency anyway; industrialization was about restoring the power of those at the top by minimizing the value and price of human laborers. This became the embedded value system of industrialism, and we see it in every aspect of the commercial landscape, then and now.

Of course industrialism wasn't sold to us as the disempowerment of workers, but as the triumph of technology. As far back as Prince Albert's Great Exhibition in 1851—the original World's Fair, really—the public was dazzled by exhibits featuring wondrous new technologies that seemed to run without any human workers at all. Attendees marveled at the mechanical looms used to make rugs in India but were never exposed to the laborers who operated them, lost fingers to them, or were displaced by them. Instead, manufacturers proudly demonstrated how their machine-produced rugs showed no evidence of human craftsmanship. Today's technology fairs, from the Consumer Electronics Show to South by Southwest, offer audiences the latest in digital gadgetry with nary a mention of the labor going into their assembly, the death of slaves mining for the "conflict minerals" they require, or the agricultural regions destroyed by the pollutants released in their production.

Let's call it the dumbwaiter effect. Remember the ingenious little hand-operated elevator Thomas Jefferson invented for his servants to deliver meals up from the basement kitchen to the dining room? Today we think of it as a time- and energy-saving technology. But for Jefferson, it had less to do with saving his kitchen staff needless walks up staircases than with distancing himself and his guests from the harsh realities of slave labor—at least while they were eating. With the dumbwaiter, food simply arrived as if from a *Star Trek* replicator. No human effort, or discomfort, needed to be recognized. This became an important strategy as the rest of human society became increasingly dependent on unseen labor.

While mass production disconnects workers from skills and the creation of value, mass marketing now disconnects workers from the people they're serving. Mass-produced products may have lost their handcrafted quality, but they made up for it with consistency. The real obstacle to their adoption was the lack of a human connection between producer and consumer. Instead of buying oats out of a barrel from Joe the local oat miller, we were supposed to buy oats in a box, shipped from hundreds of miles away. At best, the relationship with the craftsman was replaced by one with the human salesperson—who was really just a surrogate consumer at the wholesale level. Manufacturers had to substitute something for the lost human bond between consumer and producer, or supersede it where it still existed.

This is where branding came in. Putting, say, a Quaker on the box of oats gave consumers a new face to look at—and one more consistently friendly than that human miller's. Unlike the real craftsperson, the brand icon could be embedded with whatever mythology the marketer chose and forge an even deeper connection with the consumer. Of course, brand mythology had little or nothing to do with the product or its manufacture. If anything, it was a way of distracting consumers from the reality of the factory, its conditions, and its great distance away. Meanwhile, expositions and world's fairs have always celebrated the machines of industry over the humans who operated them. The Victorian exhibition displayed mechanical looms with no laborers, and the 1964 World's Fair conveyed people on moving sidewalks while showing them peopleless, automated factories. In today's computer-animated TV commercials, shiny parts fly together into a completed appliance or vehicle as if by magic. Consumers may as well be buying their cookies, cars, and computers from the machines themselves.

Finally, mass media—itself the product of industrial-age technologies, from the printing press to the satellite—gave manufacturers a way of spreading their brand mythologies across the country or around the world. Thanks to this advertising, consumers could forge relationships with brands before they had even arrived on the store shelves. That Quaker may as well have been an old friend smiling from the package. Or better.

But this last stage of industrialism came with a human price as well. Just as mass production devalued human labor, and mass marketing separated consumers from producers, mass media isolated human consumers from one another. Those fashion and perfume spots promising friends and lovers are not intended for those who already have friends and lovers. Advertisements work best on lonely individuals. So it's no coincidence that mass media tend to atomize us, creating millions of markets of one person each. That's how the television evolved from a hearth in the living room watched by the whole family to a television in each bedroom and a cable channel or YouTube stream for each person. They don't call the stuff on television "programming" for nothing—only in this case, it's humans being programmed, not machines.

In a snapshot, the transition from peer-to-peer, artisanal economic values and mechanisms to those of the industrial age looks something like this:

	ARTISANAL 1000–1300	**INDUSTRIAL 1300–1990**
Direction	•	↗
Purpose	Subsistence	Growth
Company	Family business	Chartered monopoly/ corporation
Currency	Market money (support trade)	Central currency (support banks)
Investment	Direct investment	Stock markets
Production	Handmade (manuscript)	Mass-produced (printed book)
Marketing	Human face	Brand icon
Communications	Personal contact	Mass media
Land & resources	Church commons	Colonization
Wages	Paid for value (craftsperson)	Paid for time (employee)
Scale	Local	National
Optimized for	Creation of value	Extraction of value

We'll trace each of these developments more fully, but for now what's important to notice is that each industrial innovation diminished the value of one human element after another. Identifiably crafted products, such as the manuscript, gave way to mechanically reproducible ones, such as the printed book. Instead of relating to products through the human who made them, people relate to the brand on the package, and so on. People are disconnected from the value chain. This was by design, even if the intentions behind that design have been submerged and forgotten. Remember, industrialism's primary intent was to subvert a rising middle class and their peer-to-peer market system. Merchants and craftspeople were creating value from the bottom up and threatening the passively earned income of the aristocracy. The object of the game was to get people out of the way, because they create value independently, demand compensation, and value their relationships over their purchases. And the game remains the same to this day.

Removing the human hand from industry has its obvious advantages, some of which benefit everyone in the long run. We should be grateful for the printed book for easing the spread of knowledge and ideas, even if its happy invention was part of a larger, less beneficent drive toward mechanization. Industrial processes often let us reap more, produce faster, and distribute more widely. But they do all this from a rather flawed starting place, which is why they are now reaching a point of diminishing returns.

Our measures of economic success, from corporate profits to gross national product (GNP), specifically ignore the human component of the economy. That's how an environmental disaster and its resulting cancer rates can still be considered a net positive to the economy. They require more spending on cleanup and chemo, so it's good for business as we currently define it. In less morbid examples, from corporate layoffs to tax law, we have set in place an economic system whose growth works against our own prosperity. We have lost track of the purpose of the economy.

To whose benefit? Certainly not the workers being paid less, the craftspeople whose skills are devalued, the consumers whose social ties are degraded, or the communities to whom costs are externalized. Yet we continue to optimize our businesses and our economy for growth, even as we transition toward an entirely different technological and social landscape—one with very different potentials.

This is why the leading voices today are those that still treat the emerging digital economy as Industrialism 2.0 or, as Massachusetts Institute of Technology professors Erik Brynjolfsson and Andrew McAfee put it in the title of their respected business book, *The Second Machine Age*. It's no wonder such ideas captivate the business community: for all their revolutionary bravado they are actually promising business as usual. Workers will continue to be displaced by automation, corporations will remain the major players in the economic landscape, and it's up to people to keep up with the pace of technological change if they want to survive. This is not a revolutionary vision but a reactionary one. Everything is supposed to change except the economic platform and its bias toward growth—which is probably the most arbitrary piece in the whole puzzle.[4]

Digital media and technology, more than simply giving us the means to build new machines for old purposes, offer us new outlooks on sacred truths. We gain the opportunity to reboot our economies along fundamentally different premises. Yes, we have an important new set of tools, but we are also living in what we may call a new media environment in which to make decisions about their use. As members of a newly digital society, we are learning to think about things in terms of programs and programming. Everything from the U.S. Constitution to religion to the banking apps on our iPhones can be understood as lists of commands—programs with functionality and intent.

When it comes to digital business, so far most efforts lack this depth of vision. We still tend to see digital technology as a new tool through which to scale up industrialism. So instead of mechanical looms replacing humans, it's robots and algorithms. Instead of creating distributed mechanisms to enhance the emergent peer-to-peer marketplace, we create platforms to extract value from its participants and deliver it upward. We're in a new environment but remaining true to the old growth agenda.

That's only natural. We tend to use new media in old ways, at least until we discover their innate possibilities. The first television shows were simply stage plays with a camera in the audience. The first graphical computer interfaces imitated the real-world office desktops they replaced. Likewise, our digital economy is still more in its "horseless carriage" phase than in that of the automobile—more "moving pictures" than full-fledged cinema. That is, we conceive of the digital in terms of the limits of the previous landscape rather than the potentials of the new one.

It's time to understand these potentials not as threats to business as usual but as the true promise of digital economics.

THE DIGITAL MARKETPLACE: WINNER TAKES ALL

Those of us who thought the digital marketplace was going to look something like a Burning Man festival got it wrong—at least in the short term. The distributed nature of the net, with its decentralized connectivity and ad hoc social activity, appeared to augur an equally distributed marketplace. Instead of buying everything at Walmart and watching our personal and community wealth extracted by a highly centralized corporation, we would now enter a new phase of peer-to-peer commerce. The values of the bazaar would be revived by the Internet. A new, digitally enabled, people-driven economy would dominate as industrialism's extractive growth mandate receded.

So what went wrong?

In a nutshell, we went and implemented our digital business plans without any real awareness of the deeper principles we were espousing, much less the ones we were challenging. For too many—and there's no point in name-calling here—the Internet was supposed to take care of all this. If we just built the platforms for Internet commerce, they would be inherently decentralizing, empowering, and good for mankind, as if by their very nature. We made the same old mistake of underestimating the importance of our roles as purposeful human programmers in determining how these platforms would impact the existing marketplace.

In the earliest days of networking, most of us couldn't even imagine commerce occurring on the Internet. Back in the early 1990s, the Internet was still more like a public utility than a commercial platform. Government and schools paid for it and maintained strict rules on how it could be used. It was treated much like a commons: People who wanted access to the net had to digitally sign an agreement attesting to their intentions to do nonprofit research and promising not to try to sell anything. Sending an advertisement through e-mail or posting an offer of commercial services on a Usenet group would get you kicked off the net.[5]

Those of us who were interested in digital networks and personal computing were considered freaks by the yuppies who had taken real jobs after college. Computing was seen as a West Coast phenomenon, a tool for creating graphics for Grateful Dead shows when it wasn't outputting payroll ledgers or guiding antiballistic missiles. My first book on cyberculture (originally canceled by an editor who thought the net would be "over" by my 1993 publication date) put me in contact with former 1960s celebrities such as Timothy Leary, as well as present-day counterculture enthusiasts.

I was actually in the greenroom of a CNN bureau, about to go on TV to explain "cyberspace" to Larry King, when I met Matthew Nelson, the unassuming boyfriend of my publicist. He had assembled a presentation on his laptop, with pictures, of his vision of a new way of using the Internet as a marketplace. It was a simple window, with little square icons, each containing a picture of a record album. The idea was that you'd be able to click on one of those icons, which would then fill the window with a new set of text and images. These might let you play selections from the record and ultimately click on a button that let you buy the record or CD, which could then be mailed to your home. Someday, maybe connection speeds would be fast enough for the music to be downloaded electronically to your computer. Later that afternoon, he had an appointment with an executive at AT&T, whom he hoped to convince of the promise of this new technology.

Matthew's demo helped me make my case on TV: No, Larry, the net isn't just for *Star Trek* geeks. It has a legitimate application in the marketplace. With everyone treating Internet users as if we were crazy people, getting acceptance from major corporations felt less like selling out than winning converts in powerful places. Before too long, Matthew and his brother Jonathan's company, Organic, Inc., became one of the first publicly traded Internet giants, responsible (or to blame) for not only the first e-commerce Web sites but also the first banner ad.[6]

Matthew was likely just as surprised by where this all went as I was. The information superhighway morphed into an interactive strip mall; digital technology's ability to connect people to products, facilitate payments, and track behaviors led to all sorts of new marketing and sales innovations. "Buy" buttons triggered the impulse for instant gratification, while recommendation engines personalized marketing pitches. It was commerce on crack.

With a few notable exceptions—such as eBay and Etsy—we didn't really get a return of the many-to-many marketplace or digital bazaar. No, in online commerce it's mostly a few companies selling to many, and many people selling to the very few—if anyone at all.

Take music. The best part of an online music catalogue is that it is unlimited in size. The local record store can hold only so many items in its bins. A Web site can list everything and anything, however obscure, at virtually no additional cost. A surfer in New Zealand can purchase a recording by a lute player in Norway. *Wired* editor and economist Chris Anderson called this the "long tail" of widespread digital access, which would support many more times the artists, writers, and innovators than could be supported through traditional distribution channels. His theory was that the low cost of reaching customers online would enable thousands of hitherto unpopular titles to become popular. Since the marginal cost of selling different music files was negligible, Anderson argued, the online merchants would now make a profit by "selling less of more."[7] The marketplace would become more diverse and support more creators as the long tail of former losers fattened.

Yet it turns out that's not what's happening. Instead, according to Nielsen SoundScan, a few blockbuster hits make up a greater percentage of all the music sold than ever before. In the days of physical albums and CDs, the industry rule was that about 80 percent of sales came from the top 20 percent of products on offer at that moment. That means that the bottom 80 percent still accounted for 20 percent of all sales. On iTunes today, the bottom 94 percent sell *fewer than one hundred copies each.* Just 0.00001 percent of tracks sold accounted for a sixth of all sales.[8] And these figures are roughly the same for every creative industry, from books to smartphone apps.

However we slice it, digital selling platforms exacerbate the extremes between superstars and those who sell nothing. This is because of a phenomenon called power-law dynamics. Normally, the popularity of everything from people to products is fairly evenly distributed along a bell curve (shaped like an upside-down U). That means just a few members of the population are extremely popular, most are moderately popular, and a few are utterly unpopular. Just like high school or a bookstore, most kids and books are in the hump in the middle of the bell curve.

Internet business experts expected digital platforms to flatten out that curve, giving less popular members the opportunity to connect with new friends, audiences, or consumers while also taking away some of the advantages enjoyed by the entrenched stars of radio and television. But instead of a new fatter, longer tail for formerly obscure products to thrive, we got an extraordinarily "hit heavy, skinny tail."[9] Instead of a flatter bell curve with a big "middle class" of participants, it maps out like a steep slope upward, from losers with nothing at the bottom to winners with everything at the top. This is what's meant by a power-law distribution—basically, a winner-takes-all disparity, like the infamous 1 percent.

For some reason, the original industrial-age mandate for extractive, monopolized growth was not only still in force but getting worse. Net economists were quick to defend these market dynamics as natural phenomena. "This has nothing to do with moral weakness, selling out, or any other psychological explanation," explained Clay Shirky in 2003. "The very act of choosing, spread widely enough and freely enough, creates a power-law distribution."[10] Others went on to use these naturally occurring power-law dynamics to rationalize the injustices of capitalism and increasing wealth inequality. It's just a function of increased choice—a product of human freedom itself.

What we were missing is that once again, and in spite of how social they may appear, these platforms remove living human beings from the process of selection. We are looking at numerical ratings, online reviews, and top sellers. We chat to customer service "bots" that are programmed to stoke our purchases, or people reading from automated scripts, who may as well be bots. Like most online platforms, selling sites remove the sort of human buffers and intervention that often slow things down and replace them with frictionless digital cycles that push toward extreme outcomes.

For instance, when a brick-and-mortar CD store plays a particular song on its audio system, the tune will sell more copies. But that's just one store. What happens when a subtle cue takes the form of an online recommendation on a Web site? The recommendation leads to increased sales, which then becomes a new data point that is fed back into the automated system through which more recommendations of the same or a similar sort are made. This is called positive reinforcement, but in the digital realm it's more of a feedback loop, instantaneously reinforcing itself again and again, growing and spreading. The overwhelming variety of possibilities leads us to gravitate to machine-winnowed lists, if for no other reason than to make the selection process more manageable. In turn, the more we depend on an item's popularity for discovery and selection, the more we reinforce that item's popularity and the more of a winner-takes-all landscape we create. As Harvard University researcher Anita Elberse observed,[11] the knowledge that even a single person has streamed a movie makes it more desirable than the one nobody watched yet. Two streams makes it even more popular. The outcome is inevitably winner-takes-all.

So these extreme divides between winners and losers are not simply an expression of human nature. They are an expression of one aspect of human nature, amplified by machines at the expense of all the others. In fact, it's by trying to *imitate* human, social reality that the biggest distribution platforms, from Amazon to Netflix, create and promote the distortions that lead to power-law distributions.

But the biggest and most unacknowledged factor contributing to the net's winner-takes-all effect is that these platforms are *highly centralized.* iTunes sells the music, Netflix sells the movies, and Amazon sells the books (and almost everything else). Everyone passes through the same digital turnstiles, sees the same lists and recommendations, and is subjected to the same algorithms. These monopolistic commerce platforms are not true peer-to-peer systems, and they are anything but freeing. They are growth machines—digital department stores, where the many purchase items from the few. We are all buying from the same few places and people. Continuing to do so only reinforces their position at the top, leading to more of that same centralized growth invented by the aristocracy to disempower everyone else.

Compare an Amazon, Netflix, or iTunes to a peer-to-peer platform such as eBay. Whatever one might think of eBay's corporate ambitions, its basic business model is anti-industrial, at least in that it connects buyers directly to sellers. Most sellers are real human beings with used items to unload. Although ratings matter, they are less about driving consumers to particular product choices than reassuring them of the integrity of the amateur seller whose product they have already found. The platform does take its cut—which adds up to something tremendous—but it creates a new opportunity in return. Instead of removing humans from the marketplace, eBay enhances their capacity to create and exchange value. Instead of monopolizing value, or limiting value creation to just a few players, peer-to-peer platforms *distribute* the ability to create and exchange it.

These principles can be applied intentionally by any online marketplace. For instance, Bandcamp, a music streaming and download service much like iTunes or Spotify, distinguishes itself by intentionally working against power-law dynamics. It caters to less-established underground and alternative artists, charging less than half the sales commission of its competitors. Unlike the "Top 40" emphasis of most music sites, Bandcamp eschews download counts and leaderboards in favor of a "discover" button. Users wade through their favorite genres the way they might have once flipped through the stacks at a record store.[12]

Current digital marketing dogma dictates that by promoting aimless surfing and sampling of music, a site like Bandcamp will only generate a "tyranny of choice," overwhelming users with so many options that they won't be able to pick anything at all.[13] But were people overwhelmed back in the day when they had to walk into a big record store and browse through the bins? I'm not so sure. And even if the net has trained people to accept the two or three choices at the top of popularity lists, I'm fairly confident we can regain the ability to shop. The real panic about such sites is that they emphasize human unpredictability and work against the industrial-age logic of removing people from every link in the value chain. They distribute the ability to grow.

At the very least, Bandcamp and eBay (as well as Maker's Row, Etsy, Indiegogo, and many other consciously programmed sites) prove that digital platforms don't have to lead to power-law distributions. They only end up that way because we're programming them to support our inherited strategies for mass production. We optimize for more directed consumer choice, less human intervention, volume sales, and monopoly control of a given marketplace. Moreover, we devalue the contributions of people, going so far as to ask them to spend their time and energy providing reviews, comments, and content—real value—to the corporations who own these platforms, for nothing in return.

When the expectation of free labor from consumers dovetails with the winner-takes-all lottery of the new mass marketplace, we end up with a whole lot of people working for nothing. After all, the power-law distribution writ large is really just another way of saying income disparity.

THE ECONOMY OF LIKES

So in spite of its interactive patina, the digital economy continues the industrial practice of preventing real people from participating in the growth economy—at least as its beneficiaries. We still get to work, and we still end up living and socializing on a landscape that feels much more like business than pleasure. There's just no money.

In fact, the digital landscape so effectively monopolizes economic activity that most people have almost nothing left to be extracted. That's why in order to maintain some semblance of growth, Internet companies had to find a way to monetize something other than cash from its users. Something measurable, countable, and attractive enough to shareholders to justify their real cash investment in the companies' stock.

That's right: "likes."

Social media originally appeared to be an alternative to the marketplace ethos of the dot-com era. After the dot-com boom and bust, fledgling social platforms such as Friendster, Blogger, and Myspace seemed to be offering a return to the more peer-to-peer sensibility of the early Internet. But the alternative value systems they created—likes, views, reblogs, favorites, and so on—became a new kind of currency. It's more than a mere trend in marketing styles away from broadcast advertising toward peer-to-peer social influence. It amounts to a shift in the way we value everything from entertainment and culture to consumer goods and the stock market. Likes are a new way to stoke the growth furnace.

Likes themselves are a metric of worth—and not just for teenagers gauging their social status. Real companies are valued in terms of the likes they can generate. Brands from soft drinks to automobiles check their social media traffic for upticks on a daily if not hourly basis. A multitude of sweepstakes ask consumers to do nothing more than like or retweet an ad for a chance to win cash and other prizes. According to research conducted mostly by social media companies, these "social" recommendations—particularly from trusted "friends"—mean a whole lot more than plain advertisements.

The economy of likes is most important to the social media companies themselves. At the time of its billion-dollar purchase by Facebook, Instagram had raised $57.5 million, was valued at $500 million, and had generated $0 in revenue.[14] It did, however, boast 49.6 million likes per day,[15] which has grown to 1.2 billion in the ensuing year and a half.[16]

Likewise, Tumblr netted negative $13 million the year it was purchased by Yahoo for $1.1 billion.[17] What it lost in earnings it made up for in social traffic of 900 posts/second.[18] Snapchat, a social media app with no revenue, turned down a $3 billion offer from Facebook—all for its users' 400 million daily, dissolving pings.[19]

Whether or not all that social activity will someday generate true, sustainable profits is still left to be seen. What we do know is that the likes, follows, favorites, and reposts are not as immediately valuable to the people and things being liked as they are to the companies who mine these big data troves for trends. In fact, social media companies such as Facebook now occasionally surprise Wall Street analysts by reporting revenue vastly in excess of their expectations. That's because the analysts are still thinking in terms of advertising dollars. The real revenue stream has much less to do with display ads than it does with the data that social media companies can glean from everyone's friending and liking—information that social media companies sell to big data market research firms such as Acxiom, Claritas, and Datalogix.

But if social media companies are going to maintain their growth, they must continue to generate more and more likes out of us humans. Since they can't take any more of our money, all these social media platforms must by their very nature harvest an increasingly large share of our attention, our time, and our data.

Users are only slowly coming to grips with the fact that they are not Facebook's customers but its product. Many Americans reacted in horror to the news that Facebook was conducting psychological experiments on its users.[20] But if they'd had any real awareness of how the company earns revenue, they shouldn't have been surprised at all. Facebook was simply attempting to show that the kinds of emotional contagion that occur in real life also happen online. Their social scientists proved that if people see a bunch of happy posts, they are more likely to make happy posts themselves. The controversy over the invasion of privacy or psychological manipulation may really just be displaced anxiety: we are just scared to see human emotion and action so successfully simulated and stimulated by machines. For Facebook proved that it had re-created human relationships but in the completely controlled setting of an online platform.

It's back to the medieval bazaar, except without the unpredictability of real human contact. It's a race for likes—a value system rigged from the beginning to reward one's volume of friendships over their quality. The more we depend on these numbers, the less truly social we become. While industrial-age processes simply removed human beings from the equation, these digital processes seek to *simulate* humanity through artificial social media. As digital consumers, we are no longer engaging with humans but with metrics. Life becomes one big power-law distribution.

This is just where the winners of the digital industrial economy want to keep things. Trust your "friends" and trust the numbers. Experts be damned. Reviews by *Consumer Reports,* where real scientists in expensive laboratories conduct meticulous experiments on products, are to be ignored in favor of free "peer" recommendations from strangers. *Professional* is just another way of saying *elitist,* anyway. Who needs a real review when you can just see how many people "like" something, instead? Market extremes otherwise dampened by expertise instead spin wildly out of control, while real-world professional experts, journalists, editors, and reviewers lose their jobs. An increasing number of job categories are challenged by the Internet ethos of free and open exchange among all those people on the skinny, profitless expanse of the power-law curve.

Social media companies grow at the expense of their users.

As if responding to this obvious critique of his Long Tail theory, Chris Anderson followed up with a book called *Free,* arguing that we should all give away our labor and products. In his view, creative professionals in particular should welcome the opportunity to give away their books, music, and other products because they build up demand for other services such as live performances and lectures. "Those $20k speaker fees soon add up,"[21] he explains. Of course, the only venues capable of paying those fees are corporations looking for speakers to validate their practices and reinforce the cycle of dehumanization—not schools and communities seeking help in resisting those forces.

Musicians, meanwhile, are supposed to sign "360 deals," named for the idea that they are agreeing to let a single corporation manage all their recordings, performances, merchandise, product placements, and residuals. Albums may not generate much income, but a sold-out concert can. Only now, that concert needs to be supported by a steady flow of online media and new releases.[22] According to industry expert Bob Lefsetz, musicians have to give up the quaint notion of sitting around in a studio for a year developing an album.[23] The album won't make money, anyway. Musicians' new job is to develop a constant flow of fresh content to an audience that will forget them in a few months if they don't. It's all singles, and all designed to get and stay on the iTunes list and Pandora rotation—which themselves don't generate even a fraction of the revenue musicians used to collect off album sales. That money comes only from going on the road, and only if you are a superstar.

For the rest of us, the math of "free" doesn't quite add up, unless we happen to own the platform. Those of us who wrote for the Huffington Post for years did so because we felt we were contributing to a progressive community platform. No, we were not paid in dollars, but there was a sense of solidarity in supporting a new kind of journalism, and a mutually reinforcing credibility when we all participated. When Arianna Huffington went on to sell the entire enterprise to AOL for $315 million, she did not cut her nine thousand unpaid writers in on the winnings.[24] It was as if by receiving exposure on the Web site's pages, we were already the beneficiaries of Arianna's largesse.

In reality, Arianna wasn't selling a profitable business in the traditional sense. The Huffington Post was poor in cash but rich in likes and follows. That's what she was selling. And that's what many of us are learning to sell, too. For it's not the people or their work that matter but the data their activities generate. This stuff—these likes—are not an entirely phantom metric used to fool shareholders. They are worth real money, either to brands that want to become "friends" with the fans of a particular celebrity or, better, to market researchers who want to gather data about a particular demographic.

In a landscape dominated by social media, everything begins to matter less for what it is than for how many likes it can generate—because more likes means more data to sell. The music, movies, and TV shows that entertainers create matter less to their careers than the volume of social media activity they can drum up around them. Rock videos and TV series are cast based on the number of followers a star can bring along with her. Artists and entertainers are no longer performing for human audiences so much as for the big data computers. Nursing one's Twitter or Instagram following is compulsory. Instead of taking acting lessons, the aspiring star must stir up social media attention and keep feeding users more content in order to draw out more likes from them. Given the way attention works online, this means resorting to the least-common-denominator antics: wardrobe malfunctions, sex tapes, and other usually degrading sensationalism.

Cultural judgments aside, this online social climbing leads to a strangely circular career path: creators must develop social media networks in order to "make it." But then once they've made it, the main thing they have to sell is not whatever talent they've come with but the social media network they have amassed. Yes, a famous rock star can still make money on a tour, just as a TV star gets paid for appearing on a sitcom. But these jobs are really just fodder for the bigger prize of becoming a media property oneself. Just as Arianna did.

There is a certain, if limited, empowerment in all this. A large factor in making it as a performer or even a journalist was always the ability to generate advertising revenue. In traditional media, the advertiser could dictate to TV networks and newspaper editors. If a show didn't somehow serve the advertiser, it was pulled. Likewise, the entire notion of "unbiased" journalism emerged only after national brands demanded neutral backdrops for their advertisements, so they wouldn't be accused of backing one side. By working their own social networks, creators no longer have "the man" looking over their shoulder. The new solution is to *become* "the man."

"You are your own media company," Oliver Luckett, founder of the first real social media talent and marketing agency, theAudience, explained to me when I pressed him on it. "One hundred percent. That is every single person's goal in this." Working with online celebrities from Ian Somerhalder and Steve Aoki to Russell Brand and Pitbull—people with multiple millions of followers and likes—Luckett uses a social data analysis platform to match his clients' social networks with the right brands. So if 10 percent of a TV star's million followers have also engaged with a particular shampoo or automobile brand on social media, Luckett is armed with data that can win his client a new social media endorsement. Likes for sale.[25]

Pop stars like Jay Z take it to a new level, distributing free music apps that log users' contacts, geolocation, and even phone records, all to scrape more user data,[26] which is in turn sold to advertisers and market researchers. It's as if no matter what business you're in, profit ultimately rests on your ability to glean and sell the data associated with your transactions. Even on e-commerce sites, in many cases the profitability of retail transactions pales in comparison with that of the big data they leave in their wake. Creating relationships with consumers is really just about engendering enough trust to get them to share their data assets with you.

Artists, publishers, newspapers, entertainers, and cultural producers of all sorts will have to be tuned to, if not entirely geared toward, reaching easily identified social audiences. This is not a soft science, like determining a printed magazine's audience in the old days. It's hard data on engagement. As an author, my books will be less valuable as objects for sale (people won't be paying for things like books anymore, anyway) than as the publishing tool through which I accumulate followers on social networks, whom I then sell to brands. So my books had better be brand-friendly and my audiences preselected for their data-richness. And even then I'll have to make it to the very head of the long tail to be of interest. Even social media deserves a better role in our lives and businesses than this.

The unsustainable endgame is an economy based entirely on marketing and advertising. In its currently inflated state, the entirety of advertising, marketing, public relations, and associated research still accounts for less than 5 percent of gross domestic product (GDP), by the very most generous estimates.[27] Furthermore, unscrupulous Web site owners have now learned to use robotic ad-viewing programs to juice their revenue from pay-per-click advertising. Most of these bot programs run secretly on the computers of everyday users in the form of malware, a kind of minivirus that co-opts a computer's processing power. Bots now comprise an estimated 25 percent of all online video ad viewers and 10 percent of all static display ads. In 2015, advertisers are projected to lose $6.3 billion in pay-per-click fees to these imaginary viewers.[28] Consider the irony: malware robots watch ads, monitored by automated tracking software that tailors each advertising message to suit the malbots' automated habits, in a human-free feedback loop of ever-narrowing "personalization." Nothing of value is created, but billions of dollars are made.

Eventually, social branding has to run out of fodder. As more and more markets lose all revenue potential except what they can make as social media marketing platforms, who is left to buy all this marketing and consumer data? Consumer goods like soap and potato chips may have been able to keep mainstream broadcast television alive with advertising, but they cannot support the multibillion-dollar valuations of Silicon Valley and the future of the entire digital economy.

Besides, consumers themselves are growing increasingly unwilling to play along. Many of us are actually willing to pay for the things we want—such as HBO or Netflix—rather than waste our time on free products and experiences that exist for no other purpose than to mine our data. Google's model of giving away everything in return for our looking at their ads and sharing all our data may be losing ground to Apple's "walled garden" model of paid apps and fee-for-service offerings. There's certainly room in the ecosystem for both options. We just have to hope that it's not only the wealthy who enjoy the luxury of choosing between them.

So far, the *Wall Street Journal*, which caters to businessmen on expense accounts,[29] is doing much better at extracting fees from its readers than the *New York Times*, whose paywall is derided by its largely leftist audience as vociferously as if it were a violation of the Bill of Rights' provision for a free press.[30] Those who ask for an honest day's wage for an honest day's work online are treated as the enemies of the free and open net. In that sense, the effort to hide the humans on the other side of our purchases worked all too well.

There are a few people who manage to use their accumulated reputational currency to sustain themselves—but they do so by eschewing social superstardom and the sponsorship it brings, and turning instead to their fans. Think of this strategy as the online, digital equivalent of a local brewery, only the locality isn't geographical but cultural.

For example, Amanda Palmer, a musician with a small but ardent following, found herself without a record label after the company realized she was only selling about 25,000 albums—a paltry amount by industry standards. So she turned to the crowdfunding site Kickstarter, hoping to raise $100,000 from her fan base to make another record. She ended up raising $1.2 million, from a total of just 24,800 people.[31] She was successful—and on her own terms—because she used social media to forge qualitatively strong connections with her fans instead of quantitative ones with the whole world.

Fortunately for Palmer, she enjoys doing the sorts of things required to keep that fan base feeling personally connected to her. She makes herself available to them almost 24/7, especially when on tour. Where some musicians might want to do their shows and escape to the hotel to wind down, for Palmer the show is just the beginning of a long night of dinner and dancing and conversation with fans in their homes, couch surfing, and more. So it's not a strategy that can be implemented completely online, and it doesn't leverage the net any more than it leverages the performer's time and energy.

Again, it's closer to the eBay model, with a seller connecting to her specific audience rather than trying to climb the generic leaderboard. Besides, she's not selling her social network to advertisers, so she doesn't need a massive following. She just needs enough people to pay for her music directly. She may not get rich this way (the $1.2 million went mostly to production and fulfillment), but she can live on to sing another day.

Interestingly enough, although her own fans love and support her, Palmer has been quite vociferously attacked by those who don't approve of her tactics. They argue that after receiving financial support from her fans, she should be excluded from participating in the sorts of barter and gift relationships she still enjoys on the road. How dare she enjoy the fruits of a "gift economy" of collaborators, meals, and lodgings while also asking for money for her own labor? And she's just one of many such artists accused of hypocrisy, in the screechy tones of heightened outrage that only the dehumanized anonymity of the Internet seems capable of generating.

This "hybrid" approach to making culture may be messy, but it's upsetting only when we look at it through the industrial lens of big corporations exploiting humans. As a business plan, it's inconsistent. But Amanda Palmer is not some monopoly company, or even a superstar performer exploiting her fans; she's one midlist singer trying to make a living in a winner-takes-all landscape intentionally designed to prevent her from forging real relationships or exchanging value with her listeners. Her mix of barter, money, and gift is actually much more compatible with the tangled, ambiguous nature of real human relationships and hearkens back to the best qualities of the preindustrial economy.

Digital platforms from social media to crowdfunding allow us to reclaim some of these community dynamics and apply them to our own business pursuits. Those of us who have become aware of the way some corporations exploit or hide their tactics may have a knee-jerk reaction against people who appear, at least on the surface, to be doing the same thing. But the relationships that small-business people are forging with their constituencies online are direct, transparent, and peer-to-peer; they are explicit, fee-for-service, *and* social.

They are relationships between real people.

THE BIG DATA PLAY

The value exchange between users and social networks, or fans and giant media properties, is entirely less direct and most intentionally covert.

Digital networks simulate the very same human social dynamics fueling the communities of artists like Palmer in order to generate goodwill and mass excitement for their corporate clients.

It's a one-sided, highly controlled relationship in which, invariably, the platforms and companies with which we engage learn more about us than we ever learn about them. Social marketing creates the illusion of a natural, nonmarketed groundswell of interest and, more importantly, provides marketers with a map of social connections and influences. These social graphs, as they're called in the industry, are the fundamental building blocks of big data companies' analyses.

Big data is worth more than the sum of its parts. It is the technology for solving everything from terrorism to tuberculosis, as well as the purported payoff for otherwise unprofitable tech businesses, from smartphones to video games. Like pop stars, these health, entertainment, and content "plays" will make no money on their own—but the data they can glean from their users will be gold to marketers. So they hope.

Indeed, it seems as if every startup is a "big data play." Yet when we take into account the fact that the revenue supporting big data apps must presumably come out of that same constant 5 percent of the GDP associated with marketing and advertising, it becomes clear that such a payout can't possibly come to pass. In fact, our increasing dependence on big data solutions may actually limit the growth it's supposed to be stoking.

Reducing people to manageable sets of numbers is nothing new to digital technology. It began long before digital spam, when the high cost of printing and mailing physical pieces of paper motivated marketers to limit their offerings to those homes that might actually be interested. They gathered publicly available data, such as tax records and mortgage information. They stored this information on physical notecards—one for each household—and then manually selected a range of cards to include in a mailing.

With the advent of computers, statisticians began categorizing people into increasingly sophisticated demographic and psychographic groups, giving rise to the first data-driven market research firms. With upwards of seventy different categories in which to put us, researchers at Acxiom, for example, could arm marketers with psychological profiles of their target audiences, helping them to match their pitches to the particular social aspirations of their customers.[32]

But they soon realized that their data offered more possibilities than this: it could predict our future choices. Using more sophisticated computers and methodology, researchers began connecting seemingly unrelated data points and became capable of determining who among us was about to go to college, who was probably trying to get pregnant, and who was likely to have a particular health problem. More than merely knowing our likely receptiveness to a pitch, they became capable of calculating, with alarming accuracy, what we human beings were going to do next. They had no idea why such a prediction might be true, and didn't really care. This was the beginning of what we now call big data.

What makes big data different from traditional market research is that it depends on correlations that make no outward human sense. That's the truly creepy part. Privacy is the red herring. Most people are still concerned about surveillance on the actual, specific things they are doing. That's understandable enough. So when both the NSA and corporations assure consumers that "no one is listening to your conversations" and "no one is reading your e-mail," at least we know that our content is supposedly private. But content is the least of it. As anyone working with big data knows, the content of our phone calls and e-mails means nothing in comparison with the metadata around it. What time you make a phone call, its duration, the location from which you initiated it, the places you went while you talked, and so on, all mean much more to the computers attempting to understand who you are and what you are about to do next. Facebook can derive data from how long your cursor hovers over a particular part of a Web page. Think of how many more data points there are in that single act than there are in the price of your car or the subject of your phone call.

The more data points statisticians have about you, the more data points they have to compare with those of all the other people out there: hundreds of millions of people, each with tens of thousands of data points.

Researchers don't care what any particular data point says about you—only what it reveals when compared to the corresponding data point in everyone else's profiles.

Combine this with the ability of the Web to keep track of individual users and you get a true one-to-one marketing solution. Instead of buying ads that every visitor to a Web site sees, advertisers can limit their ad spend to the browsers of their target consumers. It's the same technology that lets marketers hit us with ads for products we may have recently browsed on e-commerce sites—only now, instead of using our browsing histories, they use our big data profiles.

The same sorts of data can be used to predict the probability of almost anything—from whether a voter is likely to change political parties to whether an adolescent is likely to change sexual orientation. It has nothing to do with what they say in their e-mails about politics or sex and everything to do with the seemingly innocuous data. Big data has been shown capable of predicting when a person is about to get the flu based on their changes in messaging frequency, spelling autocorrections, and movement as tracked by GPS.[33]

For marketers looking for an edge, however, mere prediction isn't enough, and this is where they tend to get in the most trouble. Big data is simply a set of probabilities. Usually, it's hard for analysts to get more than about 80 percent certainty about a future human choice. So, for example, big data analysis may reveal that 80 percent of the people who share three particular data points are about to go on a diet. That's a pretty good indication of where to direct their ads for diet products.

But what about the other 20 percent, who may have chosen to do something other than go on a diet? They get sent messages along with everyone else, aimed at convincing them that they need to think about their weight. Feeling fat today? If they weren't already on the path to considering a diet, now they will be. And it's not even human beings making the decisions about who to send which ads—it's algorithms programmed to extract the most purchases out of consumers by exploiting their data sets. The algorithms use trial and error to see what works, iterating again and again until that 80 percent probability goes up to 90 percent. Fewer people find alternative paths as they are corralled toward the limited outcomes of their statistical profiles. Companies depending on big data must necessarily reduce the spontaneity of their customers, so that they are satisfied with what amounts to fewer available choices.

It's a digitally complexified version of the one-size-fits-all values of industrialism.

On the surface, the increase in customers for a product looks like growth. But it's a limited, zero-sum game, in which the reduction in new possibilities cuts both ways. Many of the companies I've visited have been cutting back on expensive, unpredictable research and development (R & D) and spending resources instead on big data analysis. Why ideate in an open-ended fashion, they argue, when they've already got the data on what consumers are going to want next quarter? It's virtually risk free. What they don't get is that using big data to develop new products is like looking in the rearview mirror to drive forward. All data is necessarily history. Big data doesn't tell us what a person *could* do. It tells us what a person will likely do, based on the past actions of other people.

The big rub is that invention of genuinely new products, of game changers, never comes from refining our analysis of existing consumer trends but from stoking the human ingenuity of our innovators. Without an internal source of innovation, a company loses any competitive advantage over its peers. It is only as good as the data science firm it has hired—which may be the very same one that its competitors are using. In any event, everyone's buying data from the same brokers and using essentially the same analytics techniques. The only long-term winners in this scheme are the big data firms themselves.

Paranoia just feeds the system. Becoming more suspicious of the data miners—as we do with each new leak about government spying or social media manipulation—only increases the value of data already being sold. The more restrictive we are with what we share, the more valuable it becomes and the bigger the market that can be made. We might just as easily go the other way—give away so much data that the data brokers have nothing left to sell. At least that would put them all in the same boat as the rest of us.

SHARING ECONOMICS: GETTING HUMANS BACK "ON THE BOOKS"

Digital industrialism turns human data into the new commodity. Only just like digital copies of our writing or our music, these copies of our data—our quantified selves—are taken for free. Those who market or analyze our data make money, while most everyone else is shut out. All the value we create—either directly, through our writing, music, and other online contributions, or indirectly, through the passive data trail we leave behind us—is basically "off the books." Only those with the platforms and apps to gather this data profit from it. How are human producers and consumers supposed to assert ourselves in a landscape programmed to remove us from both sides of commerce? Are humans an impediment to economic growth?

Programmer and avowed humanist Jaron Lanier thinks the answer is for us all to start participating in the game. Instead of freely providing social media sites and apps with data, which they in turn sell to analysts, we should demand to be cut in on the revenue. In the current system, we don't even have access to the data or the correlations that might be of value to us. Google and Facebook invisibly vacuum it up and then make money with it. In Lanier's vision, we would not only be able to pore through our own data, but we would get paid every time a researcher or company makes use of it.[34]

For Lanier, the net's bum deal is rooted in the way it was programmed from the beginning: links go in only one direction. Some early Internet architects envisioned a system of two-way links, in which any piece of content could be traced back to its source and the creator either credited or paid. Your Web site could link to my book, but I would know when it does. This wouldn't just help content creators; it would provide the backbone for a system through which individuals could be credited—in real money—every time their data was used to make an assessment. Or they could even be cut in on the value that was generated by the new correlation.[35] Since thousands of people might have contributed data to a single finding, another mechanism for "micropayments" would allow all the fractions of pennies to pass among the various parties, hopefully adding up to something substantial and ultimately—since almost everybody's data is valuable—the reinstatement of a middle class killed by power laws.[36] If you can't beat 'em, join 'em.

Though ingenious, Lanier's solution could actually dehumanize things even further. If we are paid chiefly for our data, then we are all performing for the machines instead of one another. We are earning money not for the ways we create value for people but for all the passive activities that happen to be data intensive. Our only value to this digital economy comes from those aspects of ourselves that can be quantified. It may solve the problem of getting a whole bunch of activity back "on the books," but to what end? So we can register some credits on a balance sheet? Must we accept "the books"—presumably, the double-entry ledger—as the fundamental operating system?

The problem with trying to get all human activity back on the books is that the books themselves are not neutral. They are artifacts of a very specific moment in human history—the beginning of the Renaissance—when the two-column ledger was instituted and everything came to be understood as a credit or a debit in a zero-sum game of capital management. Feeding more activity to the ledger simply cedes more of humanity and business alike to a growth-centric industrial model that was invented to thwart us to begin with.

That's the problem with any of the many new ways we have of earning income through previously off-the-books activities. On the one hand, they create thrilling new forms of peer-to-peer commerce. eBay lets us sell our attic junk. Web site Airbnb lets us rent out our extra bedrooms to travelers. Smartphone apps Uber and Lyft let us use our vehicles to give people rides, for money. Unlike many of the other platforms we've looked at so far, these opportunities don't lead to power-law distributions, because a car or home can be hired only by one person at a time. As long as you're listed on the network and have decent reviews, you should do as well as anyone else.

From the consumer's side, these apps are amazing. If you need a ride, you can open Uber and see a map of the area along with tiny icons for the available cars. Pick a car based on its location, the driver's ratings, and the estimated price. The driver finds you based on your own GPS location and your profile picture. Payment happens automatically, tip included. Airbnb is equally seamless. Enter a place and date and the Web site instantly renders a map with available options clearly indicated. Roll over any location to see a photo, details, and ratings for each. Book the room, and you'll find out where to meet your host or pick up the keys. Like Uber, it's unparalleled for choice and convenience.

For the providers, on the other hand, these services create a new watermark for how many of one's hours and assets should be grist for the ledger and ultimately in service of some corporation's growth. It's as if startups are out there writing algorithms to combat inefficiency and idleness by making sure that everything everyone owns is in use all the time. The platform collects its fee for putting user and provider, rider and driver, or guest and host together and enabling a new transaction where once there was none. Our assets are their new territory. Welcome to the sharing economy. Just as Lanier would have us share our data, these new companies would have us share our homes, cars, and anything else.

Only it's not really sharing; it's selling. In fact, just as there used to be an Internet that ran entirely on "shareware," there were originally free versions of these new asset-renting platforms. Couchsurfing.com created a global community of people who both give and receive space in their homes. Airbnb, its commercial successor, pitches itself the same way but operates very differently—not only do boarders pay for lodging, but the vast majority of rentals are for entire apartments. Their ads show people sharing an extra bedroom and a place at the family table, but the statistics reveal that the vast majority (87 percent) of hosts leave their homes in order to rent them.[37]

Homes become amateur hotels, as the original residents try to live off the arbitrage between the rent they pay, the rent they earn, and the cost of living somewhere other than home. Even if you are having trouble finding work in the digital economy, you no longer have an excuse for being entirely off the books. Just don't let the landlord find out what you're doing. Likewise, the amateur taxi networks of Uber and Lyft are great ways for otherwise "underemployed" vehicle owners to make a few extra bucks. There's no reason now to leave a worthwhile asset or hour off the books—even if the underemployed are really under*paid* freelancers working a whole lot of hours already. These apps are not about sharing space in a vehicle—like driving a friend to the train station—they're about monetizing unemployed people's time and stuff.

Although it currently has a valuation of over $41 billion,[38] Uber is no more a taxi service than Airbnb is a hotel chain. These are apps—beautiful ones but ultimately very simple ones—that make their money by encouraging people to engage in freelance versions of previously regulated industries. That's the real arbitrage opportunity here, and that's why local cabbies and hoteliers are up in arms. They have trained, invested, and conformed to numerous regulations to do what they do. A taxi medallion, required by law, can cost several hundred thousand dollars alone. So does a hotel license. These costs and regulations were not implemented out of spite but in order to maintain fair pricing, adequate supply, and a minimum quality of service. How can a cabbie make mandatory loan and insurance payments and compete on price against an out-of-work actor with a car, a smartphone, and a few hours to kill?

Uber, for one, well knows this. One of the company's e-mail campaigns proclaims that Uber prices are "now cheaper than a New York City taxi"—for a limited time only. It's as if the company is giving fair warning that its predatory pricing strategy is just a temporary measure designed to put regular yellow cabs out of business, the same way Walmart undercuts local retailers. This isn't simply a case of technology doing something better and cheaper. Uber's pricing power is the result not of some digital magic but of the company's immunity from medallion fees and its

$3.3 billion in venture funding.[39] It has the money to set low prices and be the last man, or entity, standing.

Yet again, human professionalism and skill is undervalued in a landscape that favors technological solutions and the power of capital over anything else. The London cabbie who knows even the most obscure backstreets can't compete against an amateur with a GPS for breadth of factual data. Under the guise of restoring a human, social, sharing element to these businesses, the crowdsharing apps actually replace skills, relationships, and local businesses with automated solutions—while a central server and the investors behind it can extract the lion's share of the revenue.

That's why the final indignity will be on the Uber drivers themselves, when they are replaced with the automatic cars currently in development by Uber investor Google. The app will orchestrate the movements of robot vehicles even more seamlessly than those driven by humans, and Uber's shareholders should do just as well—even better—in this more automated future. To them, the sharing economy is less a cultural ethos than part of a strategic transition toward more fully automated solutions. Peer-to-peer is not a means of including more people as value creators but a prelude to getting rid of them—first the skilled, fairly paid ones, and then the unskilled ones who took their places.

It's a pivot we've seen before. The Netflix DVD rental Web site offered more choice and more convenience than the brick-and-mortar video rental stores it replaced while employing far fewer people. The company even figured out how to compete with the skilled clerks of the best stores by developing an algorithmic system of personalized peer-to-peer recommendations. The higher-skilled staff jobs were replaced with unskilled labor putting DVDs in mailers. When the company became a streaming service, even those unskilled jobs were eliminated.

It's as if whenever we start down the path of trying to find an employment solution for people in a digital landscape, we end up in the same defenseless, jobless place. We can't get paid for our cultural product unless we're one of the Top 10 artists of the year. We can't get good at any job skill without its being automated by someone with a free smartphone app. The more time and assets we can get on the books, the faster they are devalued or replaced by a new technology.

More than two thirds of job losses are now the direct result of having one's function taken over by a machine. So far, these are mostly middle-class jobs, such as manufacturing, office assistance, and calculating. Commonsense advice to replaced workers was always for them to retrain and learn higher-level skills. Don't be a secretary, be the boss! Instead of doing a job that consists of entirely repeatable tasks, which will one day be carried out by a mindless machine, choose a career path that requires human ingenuity and decision making.

But now that so many businesses are using data to make decisions, many midlevel executive positions are also being automated. It's not just robots taking the jobs of warehouse workers; it's big data and analytics engines taking the jobs of stock researchers, mortgage adjusters, and marketers. We can't outrun our technologies' race upward toward higher competencies. This leads most workers to go in the other direction, down toward less-skilled jobs or less-skilled versions of traditional jobs, but those jobs are hardly immune from subsequent waves of automation.

Some business futurists envision digital workers succeeding by working on the Internet itself.* Just as Wikipedia marshaled the talents of thousands of individual people working online to create an encyclopedia, corporations are beginning to use what are called crowdsourcing platforms to engage online workers in a multitude of freelance tasks.

Again, such opportunities are touted by their proponents as part of the digital revolution. The CEO of the CrowdFlower crowdsourcing platform, Lukas Biewald, explains that these platforms are "bringing opportunities to people who never would have had them before, and we operate in a truly egalitarian fashion, where anyone who wants to can do

* Don Tapscott and Anthony D. Williams's widely read book *Wikinomics* points to Wikipedia as a new model for mass collaboration and value creation online. They go on to credit Amazon Mechanical Turk with creating valuable new opportunities for the next generation of digital workers.

microtasks, no matter their gender, nationality, or socio-economic status, and can do so in a way that is entirely of their choosing and unique to them."[40]

Crowdsourcing platforms, such as Amazon Mechanical Turk, pay people to perform tiny, repetitive tasks that computers just can't handle yet. Workers log into one of the platforms from home or an Internet café and then choose from a series of tasks on offer. They might be paid three cents each time they identify the subject of a photo, transcribe a sentence from a video lecture, or list the items in a scanned receipt. These are invariably mundane tasks—the sorts of data entry that wouldn't even exist were so many business processes not already tied to computer databases, and ones that will certainly be carried out by computers themselves sooner than later.

But for now, these tasks are the province of the click workers, a growing population of several million so far, who invisibly help computers and Web sites create the illusion of mechanical perfection. (That's why it's particularly fitting for Amazon to have named its service after the famous eighteenth-century magic trick in which a mannequin dressed as a Turk appeared to play chess. It was actually being controlled by a human chess player beneath the table.) The worker is not eliminated; he's just invisible.

For employers, it's a perfect realization of the industrial ideal: anyone can request work, do so anonymously, never meet the employee, and reject the results without ever paying. The labor force isn't simply replaceable; it's in constant flux, perpetually changing and responsible for its own training and care.

As digital labor scholar and activist Trebor Scholz has pointed out,[41] in crowdsourcing there's no minimum wage, no labor regulation, no governmental jurisdiction. Although 18 percent of workers on Amazon Mechanical Turks are full-time laborers, most of them make less than two dollars an hour. Amazon argues that the platform is all about choice and empowerment, that workers can "vote with their feet" against bad labor practices. But when even minimum-wage jobs aren't available to many workers today, they are empowered to make only one choice or none at all.

The other answer—one I've argued myself—is for displaced workers to learn code. Anyone competent in languages such as Python, Java, or even Web coding such as HTML and CSS is currently in high demand by businesses that are still just gearing up for the digital marketplace. However, as coding becomes more commonplace, particularly in developing nations such as India, we find a lot of that work being assigned piecemeal by computerized services such as Upwork to low-paid workers in digital sweatshops. This is bound to increase. The better opportunity may be to use that code literacy to develop an app or platform oneself, but this means competing against thousands of others doing the same thing in an online marketplace ruled by the same power dynamics as the digital music business.

Besides, learning code is hard, particularly for adults who don't remember their algebra and haven't been raised thinking algorithmically. Learning code well enough to be a competent programmer is even harder. Although I certainly believe that any member of our highly digital society should be familiar with how these platforms work, universal code literacy won't solve our employment crisis any more than the universal ability to read and write would result in a full-employment economy of book publishing.

It's actually worse. A single computer program written by perhaps a dozen developers can wipe out hundreds of jobs. Digital companies employ ten times fewer people per dollar earned than traditional companies.[42] Every time a company decides to relegate its computing to the cloud, it is free to release a few more IT employees. Most of the technologies we are currently developing replace or obsolesce far more employment opportunities than they create. Those that don't—technologies that require ongoing human maintenance or participation to work—are not supported by venture capital for precisely this reason. They are considered unscalable because they require more paid human employees as the business grows.

Finally, there are jobs for those willing to assist with our transition to a more computerized society. As employment counselors like to point out, self-checkout stations may have cost you your job as a supermarket cashier, but there's a new opening for that person who assists customers having trouble scanning their items at the kiosk, swiping their debit cards, or finding the SKU code for Swiss chard. It's a slightly more skilled job and may even pay better than working as a regular cashier. But it's a temporary position: soon enough, consumers will be as proficient at self-checkout as they are at getting cash from the bank machine, and the self-checkout tutor will be unnecessary. By then, digital tagging technology may have advanced to the point where people just leave stores with the items they want and get billed automatically.

For the moment, we'll need more of those specialists than we'll be able to find: mechanics to fit our current cars with robot drivers, and engineers to replace medical staff with sensors and to write software for postal drones. There will be an increase in specialized jobs before a precipitous drop. Already in China, the implementation of 3-D printing and other automated solutions is threatening hundreds of thousands of high-tech manufacturing jobs, many of which have existed for less than a decade.[43] American factories would be winning back this business but for a shortage of workers with the training necessary to run an automated factory. Still, this wealth of opportunity will likely be only temporary. Once the robots are in place, their continued upkeep and a large part of their improvement will be automated as well. Humans may have to learn to live with it.

It's a conundrum that was first articulated back in the 1940s by Norbert Wiener, the inventor of cybernetics and the feedback mechanisms that turned plain old machines into responsive, decision-making robots. Wiener understood that in order for people to remain valuable in the coming technologized economy, we were going to have to figure out what we can do—if anything—better than the technologies we have created. If not, we were going to have to figure out a way to cope in a world where robots tilled the fields. His work had influence. In the 1950s, members of the Eisenhower administration began to worry about what would come after industrialism, and by 1966 the United States convened the first (and only) sessions of the National Commission on Technology, Automation, and Economic Progress. The six volumes it published were largely ignored, but they did serve as the basis for much of Daniel Bell's highly regarded work in the 1970s about what he called the "post-industrial economy." His main recommendation was to make our technological progress less "random" and "destructive" by matching it with upgraded political institutions.[44]

Today, it's MIT's Brynjolfsson and McAfee who appear to be leading the conversation about technology's impact on the future of employment—what they call the "great decoupling." Their extensive research shows, beyond reasonable doubt, that technological progress eliminates jobs and leaves average workers worse off than they were before. "It's the great paradox of our era," Brynjolfsson explains. "Productivity is at record levels, innovation has never been faster, and yet at the same time, we have a falling median income and we have fewer jobs. People are falling behind because technology is advancing so fast and our skills and organizations aren't keeping up."[45]

However, in light of what we know about the purpose of the industrial economy, it's hard to see this great decoupling as a mere unintended consequence of digital technology. It is not a paradox but the realization of the industrial drive to remove humans from the value equation. That's the big news: the growth of an economy does not mean more jobs or prosperity for the people living in it. "I would like to be wrong," a flummoxed McAfee explained to *MIT Technology Review*, "but when all these science-fiction technologies are deployed, what will we need all the people for?"[46]

When technology increases productivity, a company has a new excuse to eliminate jobs and use the savings to reward its shareholders with dividends and stock buybacks. What would have been lost to wages is instead turned back into capital. So the middle class hollows out, and the only ones left making money are those depending on the passive returns from their investments.

Digital technology merely accelerates this process to the point where we can all see it occurring. As Thomas Piketty's historical evidence reveals, the ever-widening concentration of wealth is not self-correcting. Capital grows faster than the rest of the economy. Or, in even plainer language, those with money get richer simply because they have money. Everyone else—those who create value—gets relatively poorer. In spite of working more efficiently—or really because of it—workers get a smaller piece of the economic pie.

This income disparity is not a fact of nature or an accident of capitalism, either, but part of its central code. Technology isn't taking people's jobs; rather, the industrial business plan is continuing to repress our ability to generate wealth and create value—this time, using digital technology. In other words, the values of the industrial economy are not succumbing to digital technology; digital technology is expressing the values of the industrial economy. The recent surge in productivity, according to Piketty, has taken this to a new level, so that the difference between capital and labor—profit and wages—is getting even bigger.[47] Leading-edge digital businesses have ten times the revenue per employee as traditional businesses. Those who own the platforms, the algorithms, and the robots are the new landlords. Everybody else fights it out for the remaining jobs or tries to squeeze onto the profitable side of the inevitable power-law distribution of freelance creators.

But the beauty of living in a digital age is that the codes by which we are living—not just the computer codes but all of our laws and operating systems—become more apparent and fungible. Like time-elapsed film of a flower opening or the sun moving through the sky, the speed of digital processes helps us see cycles that may have been hidden to us before. The fact that these processes are themselves comprised of code—of rules written by people—prepares us to intervene on our own behalf.

THE UNEMPLOYMENT SOLUTION

A good programmer always begins with the question What problem are we trying to solve? So let's look at our situation from the digital perspective: Are we looking for new ways to grow the economy? Or are we trying to figure out how to get people jobs? Sure, it's a better goal than abstract, senseless, environment-depleting growth. But is it the ultimate aim here? Is this the most foundational question we can ask?

Perhaps so. Both the business and the technology press are filled with stories about how computers and robots change employment. In politics, almost any issue comes down to an argument to create jobs. War, immigration, housing, energy, budget, fiscal, and monetary policy debates all find their footing in employment for Americans: How do we get people back to work? How do we bring jobs back from overseas? How does the price of oil affect jobs? How do we raise the minimum income without its costing any jobs? How can we retrain our workforce for the jobs of tomorrow? It's as if the highest moral good and core human need is jobs.

I'm not so sure it should be. People want stuff. They want food, shelter, entertainment, medical care, a connection to others, and even a sense of purpose. But employment—a job one goes to, clocks in, does some work, clocks out, and returns home—isn't really high on the hierarchy of needs for most of us. Dare we even admit it, but who really *wants* a job? We are convinced that unemployment is necessarily a bad thing. Free-market advocates use high unemployment figures as proof that Keynesian-style government spending doesn't really move the needle. Leftists use the same figures to show that corporate capitalism has reached its endpoint: investors make money in the stock market while real people earn less income, if they can find jobs at all.

The seemingly endless "jobless recovery" makes no sense at all, particularly at a time when many of us are working longer hours as overextended freelancers or the nominally unemployed than we did when we had real jobs. It's hard to imagine how this all looks to young people just graduating college, who now chase unpaid internships with more energy than those in previous generations sought paying work.

But what if joblessness were less of a bug than a feature of the new digital economy?

We may, in fact, be reaching a stage of technological efficiency once imagined only by science-fiction writers and early cyberneticists: an era when robots really can till the fields, build our houses, pave our roads, and drive our cars. It's an era that was supposed to be accompanied by more leisure time. After all, if robots are out there plowing the fields, shouldn't the farmers get to lie back and enjoy some iced tea?

Something is standing in the way of our claiming the prosperity we have created. The toll collector whose job is replaced by an RFID "E-ZPass" doesn't reap the benefit of the new technology. When he can't find a new job, we blame him for lacking the stamina and drive to retrain himself. But even if he could, digital solutions require, on average, less than one tenth the human employees of their mechanical-age predecessors. And what new skill should he go learn? Even the experts and educators have little idea what gainful employment will look like just five years from now.*

In fact, jobs are a relatively new approach to work, historically speaking. Hourly-wage employment didn't really appear until the late Middle Ages, with the rise of the chartered corporation.[48] Craftspeople were no longer allowed to make and sell goods; they had to work for these proto-corporations instead. So people who once worked for themselves now had to travel to the cities and find what became known as "jobs." They no longer sold what they made; they sold their time—a form of indentured servitude previously known only to slaves. The invention of the mechanical clock coincided with this new understanding of labor as time and made the buying and selling of human hours standard and verifiable.

The time-is-money ethic became so embedded in our culture that putting in one's hours now feels like an essential part of life. What do you *do*? Yet jobs were not invented to give us stable identities. They were simply a part of the growth scheme: a way to monopolize the creative innovation and hard labor of the earlier free marketplace. Now that the labor is no longer needed—or is so easily accomplished by machines—must we still keep the jobs?

---

* I have been participating in the Open Society's "Future of Work" initiative, and the labor theorists, union leaders, and futurists in attendance—arguably the world's experts on the future of work—can't even agree on the parameters for defining a "job" from this point forward.

Not to work feels unethical. Even our society's favorite billionaires are "self-made," which, in a reversal of aristocratic values, lends an air of respectability to their wealth that passive inheritors now lack. But if we can separate the notion of employment from that of making a valuable contribution to society, a whole lot of new possibilities open up for us.

Our industrial capabilities have surpassed our requirements. We make more stuff than we can use, at least here in the developed world. Even middle-class Americans rent storage units for their extra stuff. Our banks are tearing down foreclosed homes in multiple U.S. states in order to prevent market values from declining.[49] Our Department of Agriculture is storing, even burning, surplus crops to stabilize prices for industrial agriculture.* There is more than enough to go around.[50] Why don't we give those houses to the homeless, or that food to the hungry?

Because they don't have jobs. Letting them just *have* stuff does not contribute to the great growth imperative.

Instead, we're supposed to think of new, useless things for these folks to make, then market those things to the rest of us, so that we go buy them, dispose of them, and then create more landfill. All in the name of growth. It's as if we expect consumers to fuel the production of unnecessary goods just so that people can put in more hours of work they'd rather not be doing. We're not looking to create jobs because we need more things. We employ people because otherwise we have no way to justify letting them share in a bounty created without their labor.

To most of us, this is just "the way things are," and to question the arrangement goes against centuries of precedent. Fortunately, all the reasons against overturning the scheme are based solely on the growth requirements of the industrial economic operating system—not on reality. Alternatives to the dehumanization scheme and its impact on work in the twenty-first century and beyond require challenging the underlying

---

* The places in the world where subsistence agriculture is no longer possible are themselves largely the victims of colonialism, global market inequities, or Western-owned factory pollution. By most responsibly derived accounts, we have more than enough bounty for the entire globe.

assumptions of this system and drawing more-direct lines between what people need and what they can provide. Here are a few possibilities, presented less as fully fleshed-out policies ready to be implemented in one nation or the other than as examples of some of the kinds of thinking we need to be able to do and some of the sacred truths we must be willing to reevaluate. Underlying them all is the implicit suggestion that our biggest challenge may be learning how to say "enough."

1. Work Less: Reduce the 40-Hour Workweek

We generally start any conversation about employment with the holy 40-hour workweek and work back from there, retrofitting the rest of our business and economic metrics to this fixed value. It's time we accept the truth: we have gotten so efficient at production that we don't really need everyone employed 40 hours a week anymore. We have to remap our time and labor in a way that's appropriate for a postindustrial society. This does not have to happen all at once, but we do have to develop a path toward less work.

Early efforts have been very promising for business and people alike. Juliet Schor, a sociologist at Boston College, believes we must overcome our fear of appearing fanciful or naïve and get on with the business of reducing work hours.[51] Her research shows that more working hours do not lead to a better economy, a better environment, or a better quality of life. Countries that have just begun instituting worktime reduction already have smaller carbon footprints than those that haven't. Schor has also shown how spending fewer hours on the job frees people to pursue the sorts of things they already do for free and that ultimately contribute even more greatly to the economy—from caring for the sick to teaching children. In the words of New Economics Foundation researcher Julia Slay, "What would the cost to your business be if your workers were never potty trained?"[52] Such value is treated as subservient to the money economy, when it is simply labor unrecorded or, as Lanier would put it, off the books.

Shortening the workweek has a profound effect on many interdependent systems. People have time to do things more slowly, such as walking to work, which uses less carbon. Shortening the workweek gives more people the opportunity to share available work, a form of engagement and participation that improves mental health and creates social bonds.[53] It also reduces overtime and work overload, both of which are statistically linked to mental illness and cancer. Other studies show that working fewer days promotes more civic and community engagement. People's perception of themselves as "citizens" and their time commitment to social issues[54] both increase.

Even in the United States, recent experiments in shortening the workweek have panned out better than expected. In 2008, Utah instituted a four-day working week for public employees by offering them the opportunity to shift from five 8-hour days to four 10-hour days. Fifty percent of the 18,000 people who participated reported that they were more productive, while a full 80 percent asked to maintain the new schedule after the experiment was over, citing benefits to their relationships, families, and general well-being. The reduction in overtime payments and absenteeism saved the state $4 million and reduced carbon emissions by 400,000 metric tons that year. And this was with no reduction in actual hours.[55] In California, Amador County workers initially protested when their worktime was reduced 20 percent, from five days to four, in order to justify a 10 percent reduction in their pay. Two years later, when they were offered the option of going back to a 40-hour workweek, 79 percent voted to stay at the reduced hours and pay.[56]

Just how strange would it be for successfully automating businesses to phase out work or at least wind down the hours? How about doing it without reducing employee participation in the profits? Not surprisingly, digital companies are some of the first to experiment with shorter weeks that don't punish employees. Treehouse, an online education startup, adopted a four-day workweek and grows by an average of 120 percent a year.[57] Productivity platform Basecamp has also instituted a four-day week because, as CEO Jason Fried explains, "when there's less time to do work, you waste less time. . . . You tend to focus on what's important."[58] The

Basecamp platform has become an industry standard in the startup community, so maybe its approach to enterprise will spread as well.

2. Rewrite the Employee-Company Contract: Share Productivity Gains

What's most important, even more important than the increased worker efficiency enjoyed by companies with shorter weeks, is the improvement in the health, well-being, and satisfaction of the human beings these companies were built to serve. While passive investors should enjoy the benefits of increasing productivity, so, too, should those who invested sweat equity. Most companies still use increased productivity as an excuse to cut jobs and then pay the savings back to the shareholders as dividends or stock buybacks. It's Corporatism 101, but ultimately a flawed, short-term approach—especially when productivity gains are spread across so many industries at once. Companies are amputating their human resources while also spoiling their own and everyone else's customer base by taking away their jobs. And all the while, digital productivity gets blamed for the obsolete business model it's accelerating.

Firms willing to consider changing previously unmovable pieces of the puzzle, such as work hours, also gain a competitive advantage in attracting and retaining the best talent. (You want to work Tuesdays, Wednesdays, and alternate Thursdays? No problem!) Moreover, keeping a reserve of available hours positions a company to take advantage of sudden bursts in activity, or a rush of new contracts, without having to hire and train new employees (only to fire them a few months later). In digital parlance, this means the company is more "resilient." It is a less brittle strategy in that it distributes the available work hours to many people instead of overemploying some and unemploying everyone else.

If, thanks to a new technology, workers become much more productive, a company doesn't have to fire a bunch of them and pass all the savings up to the shareholders. It can instead share the spoils with those workers or—if accelerated productivity outpaces demand—pay them the same salary to work fewer days.

A reduction in workdays is just one of many possible ways to contend with a paucity of available jobs. LinkedIn founder Reid Hoffman envisions digital technologies (like his networking platform) enabling people to abandon the end-to-end employment solutions of yesteryear and adopt a more temporary, improvisational approach to their careers.[59] Instead of seeking a job and then giving years to an employer in return for money, professionals will engage with companies for a specific purpose—more like a campaign. These "alliances" will last an average of eighteen months, during which a new product or division might be launched, a financial problem rectified, or a creative challenge solved. The project itself becomes part of the worker's portfolio, and the worker is engaged less as an employee than as a partner in the project.

The devil is always in the details: Isn't this a recipe for exploitation? When everyone's essentially a work for hire, what happens to the collective bargaining power once offered by labor unions? Would COBRA cover people's health insurance between engagements that might be years apart? What about pensions? Again, imagination and flexibility are required. New forms of organized labor—like the Freelancers Union—will emerge, and older, preindustrial ones like guilds will likely be retrieved. These sorts of changes don't happen overnight but incrementally and after much trial and error.

The beauty of such possibilities, from the perspective of charting a twenty-first-century career, is that they offer a glimpse of an employment path structured around the needs of real people today rather than the priorities of thirteenth-century factory owners who have long since left this realm. In nearly all these strategies, the underlying shift is away from hours served and toward value created. It's less symbolic and more real, less based in legacy systems and more grounded in current productivity. Instead of tying workers and our entire economy to the industrial-age machine, we reprogram our economy from the ground up.

It's highly digital, not in the sense that things are recorded in bits but in the way a digital society is free to reprogram its obsolete code. This is economics as an open-source proposition, where nothing is too sacred for reconsideration. We may even seek to maximize human well-being instead of continuing to devalue human contributions as a matter of course. Instead of ending up in the same logical dead ends, we are empowered to rethink our way toward entirely new constructs. We change the program to the new circumstances, rather than the other way around.

3. Surrender to Abundance: Guaranteed Minimum Income

A surplus of productivity should not be a problem. It's only troublesome in an economy in which markets are driven by scarcity alone and value is understood as something to be extracted from people rather than created for them. That's the zero-sum economic approach that sees something like, say, renewable energy as such a problem: how do we compensate for the loss of profit to be derived from digging oil out of the ground? When an economy has been based in exploiting real and artificial scarcity, the notion of a surplus of almost anything is a mortal threat. As individuals living in an economic landscape constructed over centuries to remove humans altogether, we fear being sacked the moment we are no longer necessary to our employers. As business owners, we fear technologies that render our claim to scarce resources or other competitive advantages obsolete.

We've gotten too competent to maintain this position. Rather, we must take a step outside the economic model in which we are living and accept the potentially scary truth that we have finally succeeded. In spite of our dehumanized approach (or maybe because of it), we have managed to produce enough stuff to give out a livable share to everyone as a matter of course, and for free. A whole lot of what used to be scarce is now plentiful, and between 3-D printing and other forms of distributed production, the rest of everything could turn out to be plentiful as well. We may be approaching what economic futurist Jeremy Rifkin calls "the zero marginal cost society," in which new technologies reduce the cost of everything to nearly nothing at all.[60]

In all honesty, I'm skeptical about digital technology's ability to deliver on this potential anytime soon. Technologies such as 3-D printing may make it appear as if former consumers are now manufacturing goods from scratch, but this is an illusion. Most of these printers fabricate items with plastic and resin that still has to be sourced from somewhere. Future models will allow people to make things out of metal or other materials, which will still have to be dug out of the ground and refined by someone, somewhere. If the environmental and labor footprint of a single smartphone is any indication, the true cost of 3-D printing will be anything but zero. Additionally, someone will still be manufacturing the 3-D printers themselves, and if they're at all like the printer manufacturers of today, they will be upgrading often. Just as street curbs today are littered with old monitors and ink-jet printers, tomorrow's will likely be strewn with first- and second-generation 3-D manufacturing technology. Another growth industry.

That said, I'd much rather we bump up against the limits of our technology and resourcefulness than the limits of our economic model. The last thing standing in the way of more distributed prosperity should be a paucity of imagination and courage. If we're really going to make human beings central to the business equation, we at least have to entertain the possibility that not everyone needs full-time employment in order to receive a share of the bounty. Reducing or eliminating the work requirement may prove less taxing on the economy, on the environment, and on our society.

This is not a new idea. A government-funded "guaranteed income" program has been proposed off and on since Eisenhower's administration first noticed the impending postindustrial surplus economy. Nixon pitched a version of a minimum-income subsidy as part of his Family Assistance Plan in 1969. He tried to make the argument that he was doing something other than taxing one person "so another can choose to live idly."[61] Most of Congress agreed with the concept, as articulated by New York Representative William F. Ryan:

> Guaranteed annual income is not a privilege. It should be a right to which every American is entitled. No country as affluent as ours

> can allow any citizen or his family not to have an adequate diet, not to have adequate housing, not to have adequate health services and not to have adequate educational opportunity—in short, not to be able to have a life with dignity.[62]

Congress even passed a guaranteed-income provision in 1970 by a vote of 243 to 155, but the Senate rejected the bill and others like it for fear it would make Americans more lazy. Not even the support of free-market competition advocate Milton Friedman was able to convince them otherwise. Over the next decade, Margaret Thatcher and Ronald Reagan's version of the social contract gained acceptance, stressing individual responsibility and pure market solutions to social problems. Since that time, the notion of a guaranteed income or negative income tax has sounded preposterous to most of us, a scheme disproved by the fall of the Soviet Union.

Our best data[63] suggests that to implement a negative income tax today in the United States—one capable of eliminating poverty entirely—would cost only an additional 1 percent of GNP. That's going from government spending representing 21.1 percent of GDP up to about 22.6 percent.

Critics of such plans are concerned that any guaranteed minimum income scheme will disincentivize work—particularly if there are clear cutoffs for participation, or what are known as "benefits cliffs." Some plans work against this by giving everyone the same amount of money whether or not they're poor or by focusing tax credits on children.

In any case, good evidence that minimum income plans lead to listlessness just doesn't exist. The most famous experiments, conducted back in the 1970s, did show an overall reduction in work hours when guaranteed income kicked in. But these were largely a result of people cutting back from 60-hour workweeks to 40-hour ones. Some people left work to finish high school—hardly a symptom of sloth. Most researchers were unable to find any statistically significant labor market withdrawal at all.[64] Even if the data is wrong and the naysayers are right, the upside of worker withdrawal is that it would make it easier for those still seeking work to find jobs. It would also free up the dropouts to contribute to society in all those ways that are currently not compensated.

4. Redefine Work: Getting Paid to Address Real Needs

Don't worry about the lazy getting too well rewarded. In a guaranteed-income or public work scheme, not everyone gets a mansion. We can still compete in a free market for one of those. People who want to live in luxury, buy lots of movie tickets, take vacations, or enjoy fine wines, well, they can work for a living. Not in useless jobs created simply to stoke employment, but doing the sorts of things that humans do best.

Unlike a guaranteed minimum income, "guaranteed minimum-wage public jobs" can actually redirect human effort toward the areas that need attention. As with the plans implemented for GIs returning from World War II, citizens are guaranteed jobs appropriate to their abilities that provide them with a living wage. The government sets those people to task building infrastructure or providing another civic service. Before you start seeing red, remember: unlike the socialism of the twentieth century, the motivation here is not to contend with scarcity but to contend with abundance. Although this may not have been the best approach to labor in a barely industrial Soviet Union, it may be entirely more effective at creating value for an abundant, digital-age America. Our problem is not a scarcity of toothpaste; it's finding enough consumers to keep all the toothpaste workers employed.

Working on questions of wealth inequality for the past fifty years, Oxford economist Sir Anthony B. Atkinson has gone the furthest to model these public work scenarios, albeit for the United Kingdom.[65] His empirical approach to the data and comprehensive analysis of historical patterns conclusively confirm the economic validity of more equitable distribution of wealth—as well as an increase in total wealth—through public service options. The activities of these workers, even though funded by the government, end up contributing to the wealth creation of others in ways that more than compensate the tax base for the wages they have earned.

Moreover, the money such workers earn tends to be spent back into the economy quite rapidly, recycling wealth to an extent that stock shares do not.

In a human-focused economy, there will never be a lack of need for humans. Although diagnosing and medicating people might someday be done better by computers, caring for them will not. Health-care workers, home aides, and nurses—not to mention teachers, companions, nannies, and child-care workers—are some of the least appreciated, most underpaid professionals in our society. We have been conditioned to think of these occupations as demeaning when they are really the most economically pivotal and personally rewarding ones.

These are the high-touch activities that cannot be replicated by machines. It's their necessary connection to human providers, their very unscalability, that makes them so incompatible with industrial-age values. These workers create value in real time, often one-on-one. This may make their services poor areas for growth, but they are more than sustainable professional paths for those who can perform them well and terrific entrepreneurial opportunities for those who develop the platforms, networks, and equipment to enable them.

The same goes for agriculture, textiles, and many other sectors where returning to local, human-scaled enterprise will lead to less worker exploitation and environmental damage while producing better, healthier products. Nonindustrial practices may be more labor-intensive, but they're also better for us all. For those of us used to white-collar jobs, the idea of growing vegetables or making clothes may seem like a big step backward toward more menial labor. But consider for a moment the sorts of activities the wealthiest Americans or most satisfied retirees engage in enthusiastically: brewing craft beers, knitting, and gardening. If there's really not enough work to go around and there are so many extra people to employ, we can always farm in shifts.

Those with a penchant for global conquest can still work overtime and become legendary by solving the real problems of our society: topsoil depletion, global warming, slave populations, and energy production, to name just a few. They can track the entire global supply chain of the products everyone is using and root out the parts that place an unfair labor burden upon certain people. (A low-cost smartphone that requires workers to dig for rare metals in dangerous mines is not a low-cost smartphone.) Some of these problems will be mitigated simply by taking our emphasis off this relentless quest to employ more people the old way. Once we're no longer worried about growing the economy mainly to create more jobs, we will be free to consider tackling real challenges, like the poor global distribution of crucial resources and the stultifying debt of developing nations.

Will any of this happen? Not likely very soon, especially in an environment where the game of competitive scarcity has become equated with fairness, and success within that scheme is seen as a sign of grace. From a strictly factual perspective, however, the reason we can't slow down has nothing to do with the supply of goods. We can make a whole lot less stuff—or even stop making more stuff—and still not end up waiting in 1970s' Soviet-style lines for toothpaste. Adopting policies aimed not at increasing employment but at actively *decreasing* it means challenging the assumption that the economy has to keep growing at all.

This would be a tough sell in a Congress that can't even agree to pay for last year's budget obligations, much less understand the difference between a debt ceiling and an operating deficit. Luckily, unlike Renaissance monarchs who depended on their exclusive power over lawmaking to rewrite the rules of commerce in their favor, we live in a digital landscape in which rules can be rewritten from the outside in. The industrial age may have been all about one-size-fits-all solutions, but the digital age will be about a wide range of distributed ones.

That's why we have to look at what we can do as business owners, investors, bankers, and individuals to program an economic operating system that works for people instead of against them.

Chapter Two

THE GROWTH TRAP

CORPORATIONS ARE PROGRAMS

Plants grow, people grow, even whole forests, jungles, and coral reefs grow—but eventually, they stop. This doesn't mean they're dead. They've simply reached a level of maturity where health is no longer about getting bigger but about sustaining vitality. There may be a turnover of cells, organisms, and even entire species, but the whole system learns to maintain itself over time, without the obligation to grow.

Companies deserve to work this way as well. They should be allowed to get to an appropriate size and then stay there, or even get smaller if the marketplace changes for a while. But in the current business landscape, that's just not permitted. Corporations in particular are duty bound to grow by any means necessary. For Coke, Pepsi, Exxon, and Citibank, there's no such thing as "big enough"; every aspect of their operations is geared toward meeting new growth targets perpetually. That's because, like a shark that must move in order to breathe, corporations must grow in order to survive. This requirement is in their very DNA or, better, the code we programmed into them when we invented them. Seeing how that was close to a thousand years ago, corporations have had a pretty long and successful run as the dominant business entity.

The economy we're operating in today may have been built to serve corporations, but not many of them are doing too well in the digital environment. Even the apparent winners are actually operating on borrowed time and, perhaps more to the point, borrowed money. Neither digital technology nor the corporation itself is necessarily to blame for the current predicament. Rather, it's the way the rules of corporatism, written hundreds of years ago, mesh with the rules of digital platforms we're writing today. A corporation is just a set of rules, and so is software. It's all code, and it doesn't care about people, our priorities, or our future unless we bother to program those concerns into it.

That's why it's useful—particularly in a rapidly changing media environment—to look at corporations as if they were forms of media: programs, written by people at a particular moment in history in order to accomplish specific goals. Once we have a handle on the corporate program, we'll have a much easier time understanding what happened when we plugged it into the digital economy, as well as what to do about it.

Marshall McLuhan, the godfather of media theory, liked to evaluate any medium or technology by asking four related questions about it.[1] The "tetrad," as he called it—really an updated version of Aristotle's four "causes"—went like this:

What does the medium enhance or amplify?
What does the medium make obsolete?
What does the medium retrieve that had been obsolesced earlier?
What does the medium "flip into" when pushed to the extreme?

It sounds trickier than it is. The automobile, for example, amplified speed. What did it make obsolete? The horse and buggy. It retrieved the values of knighthood—the sort of jousting and machismo we see in everything from drag races to NASCAR. And when pushed to the extreme, it actually leads to traffic jams, working against the whole point of cars to begin with. Or take the cell phone: It amplifies our mobility and freedom. It makes landlines obsolete. It retrieves conversation. And flipped to the extreme, it becomes a new kind of leash, making us constantly available and accountable to everyone.

The best part about looking at the corporation as a technology or medium is that, in the process, we remind ourselves that it didn't just emerge as a natural phenomenon. It's not as if businesses were getting so big that they evolved a corporate structure in order to keep growing properly. Quite the contrary: the corporation was invented by monarchs to stem the tide of a burgeoning middle class and its thriving new marketplace and usurp the growth they were enjoying. The fact that corporations were *invented* should alone empower us to *re*invent them to our liking.

So, then, what were corporations invented to amplify? The power of shareholders and the primacy of their capital. Feudal lords, who had lived off the labor of peasants for centuries, were getting poorer as the people began to make and trade goods with one another. The aristocracy needed a way to preserve their wealth and position in an increasingly free market. So they invented the chartered monopoly—a piece of paper with a list of rules—through which a king could grant exclusive dominion over an industry to his favorite merchant. In return, the king and other aristocrats got the right to invest in the enterprise. This way, they could use their wealth alone to make more money. Did the merchant need investors? For the most part, no.* But he made this concession in order to get the king's charter and protection. The investors were like shareholders, and the merchant was like the CEO. Except these shareholders were also the ones writing the laws of the land.

What did corporations render obsolete? They killed the local bazaar

---

* See my book *Life Inc.*, which traces the emergence of corporations in the late Middle Ages. The earliest charters granted to merchants make clear that the right to invest was a *concession* made from the merchant to the monarch. The merchant didn't need the cash and didn't need shareholders but took the investment money in return for exclusivity over a market and the protection of the king's army.

and all the peer-to-peer value creation and exchange that took place there.[2] They also worked against the marketplace's values of innovation and competition. If a company won the exclusive right to make clothing or to exploit the riches of the East Indies, then its only job was to extract value. It had no competition and no reason to innovate. We have to remember this part of the program because it's so counterintuitive: the core code of the corporate charter is to repress exchange, competition, and innovation. It was intended to extinguish the free market.

The third part of the tetrad, retrieval, is a tricky notion. It usually has a lot to do with cultural values—something from the deep past that gets rediscovered in a new form. Corporatism, by enhancing the power of the king and his ability to conduct great global enterprises, retrieved the values of empire. That's how we got the Renaissance—quite literally, the "re-nascence" or "rebirth" of the values of ancient Greece and Rome. This time around, instead of the Holy Roman Empire, we got colonialism. The colonial powers reduced places to territories and people to human resources from which to extract labor. Local values had to give way to those of the chartered corporations and the gunships protecting them.

With the power to write the laws of the territories in which they operated, corporations did very well for themselves. So, for instance, when the Dutch East India Company began harvesting cane, local islanders supplied it with rope. This became a profitable new industry for the indigenous population. The corporation, behaving true to its programming, sought to stamp out this local value creation. It requested, and won, a new law from the king outlawing rope manufacture in the East Indies by anyone but the chartered monopoly. From then on, anyone wanting to make rope had to do it as a worker or slave of the company.[3]

Likewise, in the American colonies, farmers were prohibited from selling their cotton locally. By law, all of their harvest had to be sold to the British East India Company at a fixed price. It was then shipped to England, where it was fabricated into garments by another chartered monopoly, and then shipped back to America for sale to the colonists. This was not more efficient; it was simply more extractive. The American

Revolution was fought as much against the mother company as the mother country.[4]

Finally, as the fourth part of the tetrad asks, what happens when the corporation is pushed to the extreme? What does it "flip into"? You've probably guessed this one already: a person. Even though the corporation and the industrial landscape may have worked to remove human beings from the value equation, there's nothing corporations strove for more consistently than to earn the rights and privileges of people. That's the basis of the recent Hobby Lobby case before the Supreme Court, which decided that a corporation's personhood entitles it to deny aspects of a health plan with which it morally disagrees.[5] It's also the driving force behind the Citizens United case, in which corporations were granted the right to free speech formerly reserved for humans—but not the corresponding limitations on campaign donations. And these cases all trace back to the most hard-fought battle of all, won during Lincoln's era, of corporate "personhood" itself.[6] The objective, true to the corporation's three other core commands, was to give railway corporations the same rights to land as that of its local human inhabitants. This way, people would no longer be able to object to railways' seeking right of passage through their towns or property.

Of course, the corporation becomes a person only so its primary benefactor—the investor—doesn't have to have any actual human skin in the game. The object here is for the investor, originally the king but now the shareholder, to be able to make money *with* his money. Instead of working or creating value, the investor provides capital to someone else—not a human being but a corporation—to go out and bring back returns. Further benefiting investors, the corporation accepts liability when something goes wrong. The investors' liability is limited to whatever they paid for their stock. They get to keep whatever dividends they may already have drawn or profits from the shares they have already sold. (That's where the notion of an LLC, or limited liability company, comes from.)

The function of the corporate "medium" today begins to make sense if we understand it as an expression of this original programming. This forgotten code still drives corporate behavior, angering critics and frustrating corporate boards alike. But the corporation has no choice other than to exercise the four sides of its original tetrad: extract value, squash local peer-to-peer markets, expand the empire, and seek personhood—all in order to grow pots of money, or capital.

The most successful and most loathed corporations of the last century all work this way. Walmart, for one ready example, lives by the tetrad. It extracts value from local communities, replacing their peer-to-peer economies with a single, one-way distribution point for foreign goods. Workers are paid less than they earned in their previous jobs or businesses and are often limited to part-time employment so the company can externalize the cost of health care and other benefits to local government. (Poverty rates and welfare expenses go up in regions where Walmart operates.) Understood as a medium, it amplifies the power of capital by extracting both value from labor and cash from consumers, and bringing it up and out from communities to distant shareholders.

Walmart obsolesces local trade. When it moves into a new region, it undercuts the prices of local merchants—often taking a loss on sales of locally available goods simply to put smaller merchants out of business. Even when it is not practicing predatory pricing, it can survive on lower margins by underpaying its workers and leveraging its size for discounts from its suppliers. In the long run, the store costs its consumers more in lost earnings, unemployment, a decreased local tax base, and externalized costs such as roads and pollution than it saves them in low prices.

Walmart retrieves the values of empire, where expansion is the primary aim. It has opened as many as one store a day in the United States alone.[7] The company sometimes opens two stores, ten or twenty miles apart in a new region, and keeps them both open until local merchants go out of business and new consumer patterns are established. Then it closes the less popular store, forcing those consumers to travel to the other one. In the fashion of a Roman territorial war, the advancing armies leave behind only what is necessary to maintain the region.

Finally, in its flip toward personhood, Walmart has attempted to accomplish all this with a human face—quite literally. The company adopted a version of the iconic 1970s yellow smiley face as a brand personality that the company dubbed Mr. Smiley.[8] Pressed to the near breaking point by anticorporate activists and an aggressive media, the company recently hired corporate-identity-makeover firm Lippincott to humanize its values and mission. Walmart's motto went from the utilitarian and immortal "Always Low Prices" to the much more humanistic "Save Money. Live Better." In the words of Lippincott, this "emphasizes Walmart's famous low prices while shifting the focus beyond price to the emotional benefits of shopping at Walmart. Saving money is just the beginning—with those savings, Walmart helps customers live a better life."[9] The new logo, an asterisk dubbed "the spark," is meant to evoke the twinkle of human ingenuity living within the brand itself. Walmart is alive.

Like the rest of corporate expansionism, Walmart is a success story—at least until its growth strategy reaches its limits. Thanks to the availability of new markets in China, the company might still be growing—but its stores are ultimately an extractive and not a contributive economic mechanism, taking value from the regions it conquers. Local wealth creation and exchange diminishes wherever the company's model is successfully operating.[10] It has to. The job of the company is to extract value from local communities and pay it up to investors. Its customer base, as well as its employee population, ultimately grows poorer.

It's not as if the company or its board has a choice. It must respect the wishes of its shareholders—which is growth. Walmart's main competitor, Costco, pays its workers more, hires them as full-time employees, and offers better benefits. This leads to greater employee retention, higher-skilled workers, better customer service, and arguably more favorable long-term earnings and market defensibility. Yet for years Wall Street has punished the company for violating the traditional corporate program regarding labor and has paid lower multiples for Costco's stock than for Walmart's.*[11]

---

* Lately, however, Costco's approach has been winning out. It is one of the few brick-and-mortar discount retailers whose "same-store sales" have been increasing year after year. This

Still, like Walmart, the majority of big corporations are playing a game with diminishing returns. You can extract value from a region or market segment for only so long before it has nothing left to pay with. Extractive economics is a bit like draining an aquifer faster than it can replenish itself. Yes, you end up with all the water—but after a while there's no more left to take.

That's the predicament in which corporations have found themselves after a pretty good run of global value extraction. Each time they ran into a wall, such as colonial resistance, they wrote new laws or fought new wars. Even great corporate losses, like the American Revolution, eventually became wins when laws were relaxed and corporatism was permitted to flourish once again. Over centuries of enterprise and expansion, it must have seemed like this could go on forever. There were always new continents of riches to conquer and new peoples to enslave. Even today, there are many who believe we just have to wait out this digital thing long enough to find out where the new territory for expansion awaits. We made it through disruptive technologies from steam engines and mass production to automobiles and television. Why should this time be any different?

Because, in reality, the limits of corporate expansion began to reveal themselves back in the 1950s. By the mid-twentieth century, corporations had already run out of new places to conquer, while people in places like India and Africa were beginning to push back. Even in America, consumers were finding themselves overwhelmed with purchasing choices, mortgage payments, and the other trappings of a consumer society. That's why Eisenhower and his successors turned to technology. They hoped it could become a new frontier. In essence, they were asking: could the virtual spaces of the electronic realms—TV and computers—provide new areas for corporate growth?

So far, anyway, the numbers are telling us no. Since the mid-1960s

---

may also be because Costco charges an annual membership fee to shop at its stores, which accounts for about 70 percent of its operating income, as well as increased loyalty from its member-customers.

and the explosion of electronics, telephony, and the computer chip, corporate profit over net worth has been declining. This doesn't mean that corporations have stopped making money. Profits in many sectors are still going up. But the most apparently successful companies are also sitting on more cash—real and borrowed—than ever before. Corporations have been great at extracting money from all corners of the world, but they don't really have great ways of spending or investing it. The cash does nothing but collect, like waste in a vacuum cleaner bag, almost more a liability than an asset. That's not what was supposed to happen, but it was an inevitable outcome of corporatism's unrelenting spread.

In 2009, a study initiated by economic futurists at the Deloitte Center for the Edge dubbed this "the Big Shift."[12] They anticipated the conclusion to which macroeconomists are now reluctantly coming—that an economy dominated by large corporations must eventually undergo a systemwide stagnation. As the ongoing study has discovered, although some digital technology firms, such as Apple and Amazon, are doing well in the new business landscape, "they are still only a relatively small part of the overall economy," which is losing steam over the long term.[13]

The study, which has been updated each year, researches detailed financial, productivity, and economic data on twenty thousand U.S. firms from 1965 to the present. In 2013, it found that while new technologies are giving companies the ability to do things better and more efficiently, the vast majority have been incapable of capturing the value from these new potentials. In other words, while per capita labor productivity is steadily improving, the core performance of the corporations themselves has been deteriorating for decades.

Note that the study does not evaluate firms in terms of return on investment, or ROI. It's not looking at how well investors do buying the company's stock. The metric of interest to Deloitte's business analysts is ROA—the return on *assets*, or everything the company owns and owes, from cash and real estate to debt and taxes. As John Hagel, one of the authors of the study, explained in his book *Shift Happens*, "The return on assets for U.S. companies has steadily fallen to almost one quarter of 1965 levels."[14] This means that for the past fifty years, corporate return on assets has been declining. Corporations may still be delivering more income to shareholders, but they are not doing so by making more profits.

The economists at Deloitte blame this on corporations' inability to capitalize on the windfall of efficiency and productivity bestowed on them by new technology. In their view, "the conclusion is inescapable: big hierarchical bureaucracies with legacy structures and managerial practices and short-term mindsets have not yet found a way to flourish in this new world."[15]

But the "new world" they're talking about is more than a half century old now and includes successive leaps in technology from transistors and solid-state through mainframes and integrated circuits to laptop computers, smartphones, and cloud computing. This is the world in which we all grew up. If corporations were going to find a way to flourish as they once did, shouldn't they have found it already?

What we're witnessing may be less the failure of corporations to thrive in a digital environment than the limits of the corporate model in any environment—and the acceleration of this decline with each new technological leap. Corporations have successfully captured the value that exists out there and converted it into static cash. They just don't know where to put this new money to work. Under the conditions of a free market, small businesses and individuals would then pick up the slack, creating new value. They might even be bought by big corporations, which would then grow even more. But the soil for such economic activity has itself been rendered fallow by aggressive corporate activity and regulation.

Incapable of raising the top line through organic growth, corporations turn to managerial and financial tricks to please shareholders. More often than not, this means that the corporation must cannibalize itself to deliver higher share prices or dividends. Boards incentivize CEOs to increase short-term profits by any means necessary, even if this means defunding research and development labs and personnel whose value creation may be a few years off.

It works, putting more cash on the positive side of the balance sheet temporarily. But that only makes the ROA problem worse; companies end up burdened with more unspent cash and a bigger block of dead, unproductive assets. Incapable of stoking innovation from the remaining employees, they go on a shopping spree for acquisitions—buying the growth they can't create themselves. Big pharmaceutical companies now depend almost entirely on tiny upstarts for new drugs.[16] Even digital companies that have grown too wealthy and unwieldy, such as Facebook and Google, now innovate through acquisition of startups—for which they pay a king's ransom. Google has turned itself into a holding company, Alphabet, as if to better reflect its new role as the purchaser of other firms' ideas. Standard accounting practice encourages it, because acquisitions are treated as capital expenditures, while real R & D counts as an expense against earnings. Once the new acquisition is absorbed, however, it is subjected to the same sorts of cost cutting that befell the parent. The expected "synergies" never quite pan out, which is why 80 percent of mergers and acquisitions end up reducing profit on both sides of the deal.[17]

Other companies attempt to lower expenses by outsourcing core competencies. Offshoring allows corporations to utilize workforces as they did back in the good old days of colonial exploitation. Finding employees overseas to work for almost nothing is easy. Indebted nations make the easiest targets. Forced to service their loans by exporting their local crops and resources, such countries can no longer offer subsistence farming opportunities to their citizens. So foreign multinational corporations end up with monopolies on employment and trade very similar to the kinds they enjoyed back in the 1600s. In newly industrializing nations, such as China and Singapore, former peasants migrate to the cities to become part of the manufacturing middle class—the low cost of their wages on the global market almost entirely dependent on artificially devalued currency.

Even if these inequities and manipulations could be sustained indefinitely, outsourcing is still not an enduring growth strategy. It's a way to cut corners, repeatedly, until there's nothing left at all.

Almost a decade ago, I got a call from the CEO of what he called "an American television brand," asking me to help him make his marketing more "transparent." Problem is, there are no televisions manufactured in the United States. As he admitted to me, his manufacturing, design, marketing, and fulfillment were all accomplished "out of house." So what would transparency reveal? His company was an administrative shell—a few accountants working spreadsheets of a bunch of outsourced activities. Following the corporate program may have cost less in the short run, but there was now no company left at all. The big new thing they needed to integrate into their corporate DNA was not Facebook, Twitter, or even big data but basic *competency*. They needed to find or hire some expertise, people who could innovate or who could add something to the product or brand that justified higher margins from consumers. What was the company's value-added? Their competency was not being challenged by new technology at all but by the underlying bias of corporatism away from creating any value. Besides, actually doing something well pays off in a longer-term timescale than most CEOs are incentivized to consider. And it means creating value for someone else, in the form of a good job or product.

Instead, companies look to deliver returns by lowering their costs—no matter what it means for top-line growth or long-term profitability. The CFO of an American office equipment manufacturer once proudly told me that he was going to invest over $100 million to build a factory in Vietnam. The facility would save an estimated tens of millions per annum in labor. I tried to explain to him that his calculations were based on variable geopolitical relationships, commodities prices, and exchange rates over which his company had no control. But he could only see how lowering costs would make his share price go up. The company was considered a growth stock, after all.

Sadly but predictably, the project was an abject failure. Office equipment manufacturers are not as good at global exchange arbitrage as the investment bankers watching their every move. For every company that thinks it can outsmart global capitalism by leveraging exchange rates and commodities futures, there are traders who know these markets better—and are already discounting the currencies and commodities involved (utilizing dark trading pools that office equipment manufacturers don't even know about). In the case of the factory, hedge funds neutralized the arbitrage before construction was even finished. The plant was closed just a couple of years later—before it was fully functioning—and written off as nearly a billion-dollar loss.

Besides, the smartest companies in America are already bringing their manufacturing back home. Apple, GM, and even Frito-Lay are celebrating domestic production the way that homespun brands like L.L. Bean and Ben & Jerry's used to. Beyond the halo it earns them from an employment-challenged population, it gives them an opportunity to build a culture from the inside out and to focus on core competencies for the long term rather than short-term balance-sheet maneuvers. Most of all, the reason to repatriate your competencies is to stay close to the products and processes that are the lifeblood of your work. That smell of the factory floor on your way up to the office reminds you not just of where you came from but of what is at the heart of your work and your culture. It's what you *do* for a living.

Although it's good for branding, company culture, and long-term innovation, repatriation is still understood by shareholders as a form of acquiescence to leftist grumbling for protectionist policies. What neither side gets is that no matter where a corporation is doing its business, it's *always* a foreign entity. Unemployment hawks argue against outsourcing jobs and manufacturing to China, but these processes were already outsourced. Corporations were programmed not to be part of the local community fabric but to replace those bonds with allegiance to distant, abstract brands. They were built to extract value from employees and consumers alike. Without conscious reengineering by a strong CEO, they can't bring long-term prosperity to the people and places where they operate. At best, they will create a false, temporary economy and total dependency—leaving no viable economic infrastructure once they shut down.

Corporate activity is less like a fan bringing in new air and promoting local respiration than like a vacuum sucking out the oxygen and taking it somewhere else. That's why the current predicament has been all but inevitable since the first monarch breathed synthetic life into a corporate charter. Technology may be involved with all this, but it's a mistake to point to things digital as somehow causative. Digital processes, applied to the same old tactics, simply exacerbate the same old problems. Outsourcing to robots is just another form of outsourcing.

The digital landscape does serve to make the bankruptcy of the corporate model all the more apparent. The speed and scale at which this is occurring helps us recognize that we are not in a cyclical downturn as corporations attempt to compensate for the disruptive impact of digital technology. Rather, we are in a structural breakdown, as corporatism—enhanced by digital industrial mechanisms—runs out of places from which to extract value for growth. The corporate program has reached its limits. Its function is to grow companies by turning active economic activity into static bags of capital. And in doing so, it has taken a liquid medium necessary for our economy's circulation and frozen it in corporate accounts. Farmers know to leave fields fallow or plant restorative crops so that they can repair and remineralize. Aggressive extraction leaves nothing.

From a traditional economics perspective, like that of a recent Standard & Poor's report,[18] the income disparity between people and corporations has gotten too wide. The logic used by the forecast is straightforward. The researchers broke down income into four main categories: labor, capital gains, capital income, and business income. In a healthy economy, there's a balance among these forms of income, with most people making money through labor or small-business income while a wealthy minority makes money off stock as either dividends or capital gains. If corporations convert too many assets from the working and business economies into pure capital, then the whole system seizes up for lack of fuel.

The main figure they cite, the Gini coefficient of income inequality, measures how much income has been monopolized by the shareholders at the top. A Gini coefficient of 0 would mean that everyone has the same amount of money; a coefficient of 1 means that all the income is being taken by just one person or corporation. According to Beth Ann Bovino, chief economist at S&P, once that coefficient goes above 0.4 or 0.45—where we are as of this writing—it hurts growth for everyone. "It's good for a market economy to have income inequality but to extremes, it can actually damage growth long term and make it less sustainable."[19] Bovino showed that it's not just the extreme of inequality that's to blame but the decline of labor and business income in the face of rising capital gains. Simply stated, it's harder to make money by working or creating value when the scales tip too far in favor of investors and shareholders.

In a sense, though, the aim of the original corporate program has been achieved: those who create value have been utterly subsumed by those who passively invest. But as Bovino is trying to warn us, corporate shareholders can't take this much money out of circulation without killing the goose. Those who run real businesses or, worse, work for a living end up like the musicians on the bad end of the long tail. Meanwhile, passive investors who depend on economic growth end up sitting on their bags of money, unable to find new productive investments.

That's why the S&P cut its growth forecasts for U.S. corporations, which are still flummoxed by the whole situation. Corporations saw themselves as so abstract, so foreign to any real place or market, they had no idea they were destroying the economic ecosystem on which they were themselves depending.

THE PLATFORM MONOPOLY

A corporation can't really see itself or gauge its overall contribution to the economy, much less society. It has always depended on people in order to execute its functions. No matter how much like a person the corporation became, no matter how many rights of personhood it won from Congress and the courts, it was still entirely abstract. It needed our arms, legs, mouths, and brains to function.

Digital technology, though, might finally give corporations the autonomy they need to make decisions without us, and even the bodies they need to execute their choices in the real world. What they want from us and for us is being determined right now—in most cases by corporations that are already running without fully conscious human intervention. They will soon be software running software.

No question, digital technology has created tremendous new avenues for growth. Apple, Google, Facebook, Amazon, Microsoft, and many other corporations have created new opportunities and new millionaires. But as a result of their extractive, monopolistic practices, the landscape is left with less total activity and potential for growth. The pie is smaller, or at best staying the same, but these digital businesses have managed to get bigger pieces of it—making it harder for every other corporation around, including themselves in the long term.

In large part, this is because they're still operating as if they were twentieth-century industrial corporations—only the original corporate code is now being executed by entirely more powerful and rapidly acting digital business plans. What algorithms do to the trading floor, digital business does to the economy. In the purely rational light of the computer program, a digital corporation is optimized to convert cash into share price—money and value into pure capital. Most of the people enabling this have no reason to believe it is harmful to the business landscape, much less to human beings.

At worst, argue today's generation of technopreneurs, we are undergoing a whole lot of "creative destruction." That's the process, first coined by Marx but popularized by Austrian-American economic philosopher Joseph Schumpeter,[20] through which the economy achieves a natural churn. Simply put, it's a description of how young companies with superior technologies or processes invariably unseat established ones. Old ways of doing things are replaced by better ones. There's pain, as companies go out of business and people lose jobs, but ultimately there's gain, as the new market establishes itself. Automobiles replace horses, destroying a host of horse-and-buggy-related businesses while replacing them with auto shops and gas stations. Portraiture is replaced by photography, and in turn Kodak, the dominant photography company, is brought down by the advent of digital cameras and smartphones. Independent bookshops are destroyed by superstores such as Barnes & Noble or Borders, which are themselves destroyed by Amazon.

In other words, this activity may be destructive to the companies or categories that die, but opportunities for new enterprises are created in the process. It sounds really promising on the surface, more like the young replacing the old or a more developed species replacing a weaker one. As Schumpeter suggests, it's just another form of evolution.

This rationale has been enough to keep most thoughtful Silicon Valley entrepreneurs from worrying too hard about the repercussions of their actions. After all, digital corporations will necessarily carry out corporate code better than their predecessors. They apply the engineer's logic to every situation or choice and always optimize for the best and most defensible outcomes. For example, last century's retailers mailed out catalogues and then used sales feedback to adjust the offerings for the next quarter. A digital company will A/B test its Web page, display ad, or online catalogue in real time. Every interaction is a test of a bigger/smaller font, a higher/lower price, friendly/formal language, and so on. The thousandth time a page is rendered, it has evolved into a much better selling mechanism. Digital is better.

Each and every choice and process can be made more efficient, more responsive to market conditions, and more persuasive to users. And why shouldn't companies optimize for victory? Whether it's MOOCs replacing in-person college courses, Web sites replacing stores, apps replacing newspapers, or streaming MP3s replacing radio, it's only creative destruction. Either get with the program or get run over by it.

That's the sanguine interpretation of creative destruction, held by the winners since industrialism began. If you're the unhappy victim of a plant closing, the resident of an abandoned community, or the owner of an undercut small business, you are an unfortunate but necessary sacrifice to business innovation and free-market competition. Free-market advocates celebrate creative destruction as the way that scrappy young upstarts come and unseat the most powerful companies on the block. But Schumpeter also suggested that each new winner takes over its sector in a much more complete way than its predecessors, potentially destroying more businesses and opportunities than it creates—certainly in the short term. It's like big fish swallowing up smaller ones until only a few really big fish remain. And with enough influence, those big fish can change the rules and further disadvantage those who would rise up to eat them.

Take the toy industry, which grew highly consolidated through the 1980s and 1990s. The top four global companies each have revenues over $4 billion,[21] after which a long tail of much smaller players begins. In 2007, after thousands of toys manufactured in China were recalled for having lead paint, the four industry leaders worked with U.S. government agencies to develop new regulations—all in the name of protecting children. The testing protocols they developed, however, cost over $40,000 per product, which made sense only for high-volume manufacturers.[22] In spite of their protests, independent toy makers were not even invited to the table. Craftspeople making toys by hand or in smaller runs had no way of complying with the testing process and were forced out of business—even though they weren't the companies outsourcing their production to begin with. Under the guise of consumer protection, the incumbents create regulations to entrench their monopolies.

Creative destruction accelerates whenever there's a major new technology capable of fostering entrepreneurial activity, so the fact that we're seeing so much churn right now shouldn't surprise us. Nor should it upset us, at least not if we take Schumpeter to heart and accept that without pain in the form of lost employment and social destruction, we won't get gain in the form of new markets for capital. But the entrepreneurs fomenting today's upheavals appear more aware than their predecessors of how to create monopolies, leverage networks, and exploit their technological advantages—even without a government to manipulate. The digital difference is that monopoly-favoring regulation needn't occur at the political level when it can be embedded in the operating systems themselves.

Uber, as we've seen, means to be the creative destroyer of the current taxi industry. It bills itself as a way of connecting drivers and passengers. According to this way of thinking, it is primarily a platform and payment system, not a taxi or limousine service. Passengers register their credit card with Uber, which sets prices, charges payment, takes its 20 percent cut, and pays the driver.[23]

By calling itself a platform rather than a taxi dispatcher, Uber has been able to work in a regulatory gray area that slashes overhead while inflating revenue. Unlike traditional, regulated livery services, Uber is under no obligation to the public good, freeing the company to implement "surge pricing" during peak use periods, as it did during Hurricane Sandy and other disasters—a practice indistinguishable from price gouging.[24] Uber also claims that its status as a mere platform significantly reduces its responsibilities to its drivers. This issue is still being hashed out in the courts and city councils, but relative to traditional livery services, the difference is clear: Uber drivers take on greater personal liability than any driver working for a legitimate, licensed cab company. When an Uber driver, in between passengers, struck and killed a six-year-old girl, Uber claimed no liability. A traditional livery service must legally assume liability for all on-the-clock drivers, whether they're currently transporting a passenger or not.[25]

This is how Uber can be valued at over $18 billion while many of its drivers make below minimum wage after expenses. Meanwhile, the company's path to success involves destroying the dozens or hundreds of independent taxi companies in the markets it serves. On the surface, it's the creative destruction of centralized taxi commissions and bureaucracy. The result, however, is the elimination of independently operating businesses and their replacement with a single platform. Former business owners become Uber's unprotected contractors.* Market pricing and competition are replaced by a monopoly's algorithmic price-fixing.

Creative destruction? Perhaps—but with a twist: the new businesses of the digital era aren't stand-alone companies like stores or manufacturers but, as they say, entire *platforms*. This makes them capable of

---

* As of this writing, lawsuits against the company contesting the independent-contractor status of its drivers are under way. Uber is objecting to the notion that it is an employer or anything more than a neutral platform enabling the business dealings between individuals.

reconfiguring their whole sectors almost overnight. They aren't just the operators—they are the environment.

To become an entire environment, however, a platform must win a rather complete monopoly of its sector. Uber can't leverage anything if it's just one of several competing ride-sharing apps. That's why the company must behave so aggressively. Uber's rival, Lyft, documented over 5,000 canceled calls made to its drivers by Uber recruiters, allegedly in an effort to get drivers to change platforms.[26] It's not that there's too little market share to go around; it's that Uber doesn't mean to remain a taxi-hailing application. In order to become our delivery service, errand runner, and default app for every other transportation-related function, Uber first has to own ride-sharing completely. Only then can it exercise the same sort of command as the chartered monopolies on whose code these modern digital corporations are still running.

Union Square Ventures founder Fred Wilson worries aloud on his company blog that digital entrepreneurs are more focused on creating monopolies and extracting value than they are on realizing the Internet's potential to promote value creation by many players. Wilson is excited about the possibility of new platforms that allow new sorts of exchange, "but," he says, "there is another aspect to the Internet that is not so comforting. And that is that the Internet is a network and the dominant platforms enjoy network effects that, over time, lead to dominant monopolies."[27] The fact that digital companies can build platform monopolies brings creative destruction to a whole new level.

Amazon provides the clearest example of traditional corporate values amplified through a digital platform monopoly. As the *New York Times* explained, "At first, those in the publishing business considered it a cute toy (you could see a book's exact sales ranking!) and a useful counterweight to Barnes & Noble and Borders, chains willing to throw their weight around. Now Borders is dead, Barnes & Noble is weak and Amazon owns the publishing platform of the digital era."[28]

It all started so innocently. With Amazon, everyone got equal footing, so small publishers could more effectively compete against the majors. No more battles over getting B&N to stock your book; this Web site sold everything to everyone. Consumers could find what they wanted more easily, read peer recommendations, and feel assured of getting the best price. Authors and others with Web sites could become Amazon Associates and make a little money for recommending books through links. As the company grew, its catalogue became a replacement for Books in Print—the industry's original title index—and its rankings became the new best-seller list. Up to that point, it appeared that an entire industry had been cracked open and democratized, thanks to the disruptive power of the Internet.

Of course, with hindsight, we now see that Amazon is less a bookseller than a business plan. As *Forbes* put it, only half admiringly, "Unlike the other big companies that symbolize our times—Google, Apple, Facebook, Microsoft—Amazon did not rise to power by inventing a new product or service. It came to power by systematically taking down an entire existing industry."[29] In all of the company's moves, in each of the ways it leverages its platform monopoly, we see the digital activation of the earliest tenets of corporatism.

Amazon amplifies the power of central authorities. It first appeared that it would empower the independent publisher by giving everyone a place on its infinite shelf space. But it eventually grew into the center of the publishing universe. Everyone is the same size—tiny—compared to the platform on which they sell and interact. Amazon sets the prices, the terms, the technologies, the copy protection, the privacy of readers . . . everything.

Nonparticipation is not a real option. Publishers and authors must decide whether to submit to its terms and pricing or go it utterly alone. Resistance to any of the platform's demands results in Amazon pulling the rebellious publisher's books from search results. That's a form of bullying available only to a near-monopoly like Amazon, whose selling platform also amounts to the Wikipedia of publishing. Yes, publishing information is available elsewhere, but Amazon has become a default—as Google is to search.

Amazon's ubiquity lets it demand better pricing from publishers, which are themselves struggling to survive in the new scheme. The company doesn't particularly care, as it will do better if publishers fold and it can distribute authors' works directly. In the fashion of the Dutch East India Company making its own rope, Amazon sees any service provided by an outsider as a profit opportunity to absorb: printing, publishing, e-books, readers, tablets, and even smartphones, streaming media, and movie studios. It has even reached ubiquity in Web-based cloud services—the platform on which all the other platforms, from Netflix to Airbnb, are based. It leverages and defends its position by having companies conform to its own proprietary software configurations. Once you're in, it's a bit like putting all your music into iTunes; good luck getting it all out again.

Amazon obsolesces the peer-to-peer marketplace through its very centrality. At first, services such as Amazon Associates and Fulfillment by Amazon appeared to be a boon for peer-to-peer activity, giving anyone the ability to post listings and sell new or used books or other merchandise to anyone else. By posting used-book offerings in the very same display as a new-book listing, Amazon undercut its own book sales as well as those of its original vendors—the authors and publishers who are now dependent and powerless. This doesn't matter to Amazon, though, because the books are just a loss leader for this bigger prize: ownership of the marketplace itself.

People selling to one another on the Amazon platform are not in a true peer-to-peer marketplace. They are not connecting directly; they are each connecting to and through the product listings on a centralized server. The ability of the net to make true point-to-point connections is not being enhanced; instead, both parties are simply interacting with a Web site that cares about nothing more than extracting a percentage of the transaction and becoming the only venue where such transactions can happen.

Amazon retrieves the spirit of empire by colonizing not just verticals within its own category but horizontals in everyone else's. It first established a platform monopoly in books by selling books *at a loss*, in the manner of Walmart using its ample war chest of capital to undercut local stores. A simple loyalty perk like free shipping was eventually revealed to be the ever-expanding, increasingly sticky Amazon Prime. Amazon then leveraged its monopoly in books and free shipping to develop monopolies in other verticals, beginning with home electronics (bankrupting Circuit City and Best Buy), and then every other link in the physical and virtual fulfillment chain, from shoes and food to music and videos.

Finally, Amazon flips into personhood by reversing the traditional relationship between people and machines. Amazon's patented recommendation engines attempt to drive our human selection process. Amazon Mechanical Turks gave computers the ability to mete out repetitive tasks to legions of human drones. The computers did the thinking and choosing; the people pointed and clicked as they were instructed or induced to do.

Neither Amazon nor its founder, Jeff Bezos, is slipping to new lows here. The company is simply operating true to the core program of corporatism, expressed through new digital means. Amazingly, as of this writing, anyway, Amazon itself operates at a loss. Its share price is the only thing that's going up, currently sustaining a market cap of over $150 billion.[30] But in a deeper sense, this means the corporate program is working perfectly: all the value is being accumulated in the investors' shares, which are still going up.

Amazon isn't really a new sort of company so much as a very old sort of company. It is leveraging digital platforms the way colonial powers once leveraged their exclusive shipping routes to the New World. (Both even have pirates to watch out for!) That's why none of this is ever about bringing more value to people or—heaven forbid—helping people create and exchange value on their own. Digitizing the corporation simply affords it ever more efficient and compelling ways to extract what remaining value people and places have to offer.

This is why we shouldn't be so surprised that most of the strides in artificial intelligence have a whole lot more to do with computing power than with human potential. Projects such as IBM's Watson or Google's Machine Learning lab are not augmenting human intelligence so much as creating systems that think for themselves. With every keystroke and mouse click we make, their algorithms learn more about us while simultaneously becoming more complex than we—or anyone—can comprehend. They are getting smarter while we humans are getting relatively, or perhaps absolutely, dumber.

Our machines slowly learn how to manipulate us. It's a field now called captology: the study of how computers and interfaces can influence human behavior. At Stanford, where it is taught in the computer science department, captology is pitched as a way of helping people live happier, healthier lives. The examples of captology at work on its Web site include rewarding people with pleasing graphics and sounds when they balance their checkbooks online or reach a target weight as measured by a digital scale. But the real market for thinking technologies is corporations looking for ever more powerful ways to compel humans to act.

Imagine a world where online purchases are stoked by photos, colors, and appeals assembled by algorithm and fine-tuned to our individual profiles. Or apps that make sad gong sounds or display little frowns when you decide to turn them off. Well, they're already here. As our machines become more intelligent, they will become better actors, tugging at our heartstrings in all the right ways. This is not because they're alive or have feelings but because they've succeeded in carrying out their original corporate programs.

Many computer scientists and technology philosophers look forward to the day when our computers surpass human intelligence—what they call the "singularity." They hope that computers will simultaneously achieve a form of consciousness as well and carry the human project into the future even after we're all gone. That would definitely count as the ultimate flip of corporations into personhood. Sure, a few favorable Supreme Court rulings have helped, but the best enabler of corporate evolution toward personhood is this newfound digital ability corporations have to think for themselves. As the corporate program fully integrates with digital technology, it's no wonder our biggest corporate conglomerates make artificial intelligence their highest priority.

Although computers may never become conscious, they will certainly be smart, and their ability to iterate rapidly will prove challenging for human beings on both sides of the corporate equation. At that point, who is left to exercise any human intervention in business? The people running corporations can no longer credibly claim that they are merely responding to consumer demand, since consumer demand will be largely determined by smart machines. And those machines will simply be running the original and unchallenged corporate program as best they can, carrying out a thirteenth-century template for converting value into capital, replacing human agency with the corporate agenda, and usurping organic growth by creators in favor of monopolized extraction by established players.

Most significantly, they will do it faster and better every day, learning and improving with every action. The programs with which we carry out our daily business, from online shopping to employment platforms, are all optimized to accelerate. This is why these business plans are spinning out of control, toward the extremes we're now witnessing. We took a program that used to require human actors to execute and put it on a digital platform.

Ironically but irrefutably, *this is not good for business.* As more value is sucked out of the economy and frozen in corporate storage, companies' return on assets erodes even further. As corporate algorithms battle one another for platform monopolies, the extraction of value and opportunity from the real economy worsens. An app swallows an industry and has nothing to show for it but shares of stock with no earnings. On a digital landscape running only corporate code, corporations themselves end up in the same predicament as musicians and everyone else: a couple of winners take it all while everyone else gets nothing. Making matters worse, remember, in a successful corporate environment total economic activity *decreases* as money is sucked up into share value.

It's as if the business world is morphing into a video game. We can only wonder who the eventual winner of the growth game will be as the Gini number creeps upward toward one. Sergey Brin, Mark Zuckerberg,

Jeff Bezos . . . ? They're playing in a winner-takes-all competition. Google is trying to leverage its platform monopoly to become a shopping platform, Facebook is leveraging its monopoly in social media to become an advertising service, and Amazon is leveraging its store to become a cloud service.

In the corporate program, there's only room for one.

RECODING THE CORPORATION

CEOs are coming to recognize digital industrialism's diminishing returns. Some are scrambling to extract what remaining value they can before the inevitable crash. Others, however, are coming to realize that companies in a digital landscape don't have to approach their markets with scorched-earth determination or with the biases of traditional corporatism operating as their core code. They don't have to grow in order to stay alive, or achieve absolute monopolies in order to achieve their goals. Digital companies in particular have the ability to rethink some of these assumptions and rapidly deploy new approaches.

For example, as an alternative to investing in the platform monopolies favored by most of today's venture capitalists, Fred Wilson has invested in Uber competitor Sidecar, which he argues "has built a true open marketplace for ridesharing."[31] Sidecar does not offer the extreme convenience of Uber, but it's not really geared toward increasing the efficiency of business travelers. It's more of a peer-to-peer ridesharing app, through which passengers book lifts from drivers usually in advance. The app lets passengers connect with drivers and then gets out of the way, emphasizing those human-to-human connections over the primacy of its own platform. In contrast to Uber's centralized price-fixing and opportunistic gouging,[32] Sidecar asks drivers to set their own prices by negotiating with riders within the application. Sidecar facilitates a decentralized free market.

The app has been configured to transcend the traditional biases of the corporation against real-world human connection. It's a more social rideshare program, not a gray-area, unregulated taxi service, so it's not competing head-to-head against full-time livery drivers. Uber surely wins in an always-on world where agility is the key to success. Sidecar wins in a slightly slower world where riders plan ahead, giving them the added luxury of being choosier about price, driver, and amenities. Think Grandma's weekly drive to the hairdresser or grocery store. She might even find a driver she likes and book him regularly. The fabric of local connections begins to assert itself. For their part, the part-time drivers are less reliant on the platform for income and thus less likely to accept a pittance for their services.

In some sense, these apps are each configured to their respective visions of the world. Uber's is a corporate-driven world where speed and convenience trump socializing and planning. It exploits a platform monopoly to extract value from its users, while Sidecar attempts to help its users create and exchange value in a new way. Uber's reviews and other capabilities are worth more to us in an anonymous landscape, where we are depending on this information to judge one another. Sidecar depends more on its users' finding favorite drivers, engaging in repeat business, and setting up regular schedules.

At the very least, apps consciously engineered to promote the net's connective potential may just disrupt the disrupters. As Wilson explains, "We think it's more likely that [a] true peer marketplace will keep Uber honest than the legacy fleets of limos and taxis that are fighting for their life against Uber right now."[33] For now, these peer-to-peer alternatives are limited to local scenarios like finding a lift to work or a babysitter for one's kids. Those are places where the regional and human elements still offer a competitive advantage.

In the venture capital game, however, Uber clearly wins. Sidecar has taken in about $20 million, while Uber is worth closer to $20 billion. The bigger, centralized solutions offered by corporations with traditional, extractive, and monopolistic strategies are more attractive to investors, who are themselves betting on winner-takes-all outcomes.

But companies that can reach profitability through real-world revenues and then pay back those dividends to employees and other stakeholders may actually be in a much more defensible position moving forward. For them, growth is not a requirement but a happy side effect of doing good business. Becoming bigger could make the businesses' owners a bit wealthier, but for the most part it just means more individuals will be able to participate in value creation. The winnings are not accumulated in stock. They are distributed in salaries, dividends, and services.

The next wave of digital businesses may just look more like Sidecar than Uber. Given the limits of the marketplace to provide more capital to the investor class, it may *have* to. Instead of striving for platform monopolies that mirror the monopoly charters of the earliest corporations, these more distributively conceived digital companies are looking to the peer-to-peer architecture of digital networks for inspiration. They end up with models less dependent on establishing and enforcing monopolies and less encumbered by the growth imperative. In almost every sector, from health to clothing to 3-D printing, distributive alternatives are challenging the platform monopolies.

Take education. The leading digital platform for college class management is called Blackboard. With over 17,000 schools under contract, it is on its way to platform monopoly in this sector and is already worth a couple of billion dollars in private equity.[34]

The Blackboard system runs as a big, centralized server that controls who has access to what. It is a one-stop shop for everything associated with teaching and learning, from group e-mails to video assignments to grading. Everything happens through the platform, which gives Blackboard the ability to take control of more and more classroom assets as it is used. For example, it's easy to e-mail a student or an entire class through Blackboard but impossible to find a student's e-mail address in order to communicate through any other means. The more education and administration Blackboard subsumes, the more dependent everyone becomes on it, the less reversible the decision to use it, and the more easily the company can leverage its control to upsell something else. Mention the name Blackboard to an educator, and you'll get an instant tirade on the pitfalls of centralized platforms. In the true spirit of a platform monopoly, the company attempted to patent the entire concept of connecting Web-based tools to create an interconnected, universitywide course management system[35] and sued its lone competitor for infringement.

In stark contrast to Blackboard's winner-takes-all colonization of the education space, a Knight Foundation–funded startup called Known[36] attempts to fulfill the same classroom functions without any centralization at all. It models itself after what has become known as the "open Web"—a series of protocols through which people, Web sites, and applications can interact directly. Back before Facebook, for example, people managed their own Web sites on separate servers. Facebook brings all those Web pages together into one place, but users must surrender authority over their information, ownership of their data, and even the ability to reach everyone in their own network of friends (unless they pay a premium to do so). Open Web advocates seek to create services and applications that restore the peer-to-peer quality of individual Web sites. They make their own platforms as "thin" as possible, so that users can keep their own data and feed one another stuff. Think of it as a bit more like a newsfeed or e-mail than Facebook or a newspaper Web site. When there's no central repository, there's much less of a drive toward platform monopoly or the creation of absolute dependence on a single entity.

The Known project is really just a set of open-source components through which teachers and students can subscribe and publish to one another's Web pages. Teachers can make pages for their classes, then provide access to enrolled students. Those students subscribe to the Web pages of classes in which they are enrolled and receive the appropriate feeds on their own Web pages. Known provides server space for those who want or need it, but it's not at all mandatory. The real product here is not a platform but a set of protocols through which all these little independent Web pages can feed information—assignments, papers, and so on—to one another. Jane writes a paper, publishes it on her private page, and then "syndicates" it to her teacher or her whole class, depending on what she wants. Those people then see the paper on their own Web pages.

While Known can charge for server space or a fully supported version of its tools, its whole system is open source and running on open application programming interfaces, or APIs, which let anyone incorporate the apps into their own systems. The company makes money by charging for what it provides to its users, not by selling shares to speculators. As what amounts to a service company, Known will make profits that are proportionate to how many people it is serving, but it will never be able to justify the valuations of a closed platform like Blackboard. Then again, as long as companies like Known don't accept too much venture capital, they will never have to justify those sorts of valuations.

The truly successful scalable company in the digital economy may not be the one that can grow infinitely but the one that can prosper on any scale, large or small. Learning to scale down as well as up may just be the key to longevity in a moment like this, when the original corporate game plan for perpetual growth appears to have reached its limits. That's a lot easier for a new digital company to do from scratch than it is for a major corporation to pivot toward after a hundred years of pursuing growth.

It's time for the Fortune 500 company to act like a river reed instead of a mighty oak. But how?

THE STEADY-STATE ENTERPRISE

That's the question I've been getting from CEOs for the past ten years or so, and it's the main reason I chose to write this book in the first place: they're ready to listen. They don't call me up asking, "What do I do if I can't grow my company anymore?" It's usually after I've given a speech about digital economics and the possibilities for a more sustainable approach to enterprise that one of them asks me to lunch or offers me a ride to my hotel.

We talk about technology, nature, and the future but invariably—just as we're saying good-bye—the real questions emerge. *How do I transition from a postwar growth corporation to what we really are? And how do I tell my shareholders the news?*

They're always relieved to know that the steady decline in profit over net worth they've been suffering is not their own secret problem but a widespread symptom of this period in economic and technological history. The only lingering question is whether it's simply a cycle repeating itself or a unique and unprecedented challenge to our economic operating system. Although they would be the last to see it this way, even the corporations would be better off if things are happening truly differently this time out.

According to political economist Carlota Perez, who has conducted the most comprehensive analysis of how entire economies respond to technological revolutions,[37] we have passed this way before. In every instance so far of a major technological revolution—whether the steam engine, electricity, the automobile, or television—we have gone through the same five phases.

In the first phase, *maturity*, established companies from the previous technological revolution plant the seeds for a new paradigm. Electric companies invested in the first radio and television companies; Xerox, an office machines company, invested in research for the first computer user interfaces; Kodak developed the first digital cameras. Companies use their capital to invest in the technologies and industries that eventually replace them. The next phase, called *irruption*, is the technological breakthrough itself, and the disruption of the previous technology as well as the industry that built up around it. The automobile disrupts the horse trade, TV disrupts radio, the Internet disrupts TV, and so on. Next comes *frenzy*, when we see the formation of speculative bubbles, increasing unemployment, and the beginning of unrest. That's likely where we are today, with the absurd valuation of every remotely plausible new platform monopoly, as well as the joblessness and upset that the successful ones generate: cabdrivers protesting Uber, hotel workers complaining about Airbnb, and

San Francisco residents throwing rocks at Google buses over inflated rent prices.

Then the stock market bubble pops. Perez sees that as the *turning point*, when wealth disparity between the winners and losers reaches an extreme, civil unrest reaches a peak, and government is forced to act through regulation. For instance, the irruption and frenzy phases of automobiles and mass-produced appliances led to the Roaring Twenties and eventually the 1929 crash. So government and even industry begrudgingly supported the New Deal and the welfare state. Only then followed the more stable, regulated period—a golden age—when the middle class actually got to benefit from industrial technologies. This period, what Perez calls *synergy*, leads to a wider assortment of industries that support the original technological revolution but in ways that give more people access. In Perez's examples, driving schools supported the widespread adoption of automobiles, while new import and export industries extended the benefits of canals. As such industries grow and mature, they form the basis for the next round of technological innovation.

Without government intervention and initially painful regulations, however, these revolutions would never have made it through the turning point to a golden age. The industry-driven American dream of an automobile and a home in the suburbs was made possible by the welfare state. The GI Bill provided the down payments for homes, and unemployment insurance allowed workers to keep up mortgage and car payments between jobs. Without such welfare accommodations, cars would have been repossessed and homes foreclosed on whenever a factory paused production and laid off workers. This would have crippled the automobile and homebuilding industries, which were then still a backbone of the American industrial economy.

But we've seen extreme wealth disparity as well as two stock market crashes so far—the dot-com crash of 2000 and the digital finance crash of 2007—and no sign of a widespread populist cry for more regulation or a welfare state. A popular uprising for government intervention seems unlikely when current populists, such as the Tea Party, see government as the cause of, rather than the fix for, the inequities they are suffering. Even a government-supervised marketplace of corporate health insurance plans is understood by many as a form of capitalism-killing socialism. Meanwhile, progressive populists (such as the members of the Occupy Wall Street movement) went from demanding government reform to simply taking care of issues themselves, through mutual-aid efforts and debt buy-back programs. Things would have to get much worse, I fear, for today's crop of populists from either side of the political spectrum to seek redress in the form of government intervention. And then we would need an entirely more functional Congress in order to execute any genuine reform.

Besides, even if we somehow made it through the frenzy, there's no great opportunity for a "synergy" of new ancillary businesses waiting on the other side of the crisis. The young companies of today's technological revolution don't require as many employees or support industries as those of the past. That's why our predicament is not so much a case of creative destruction as one of *destructive* destruction. Real businesses and profitable interactions are destroyed, and in their place comes a corporation that does not even have an operating profit. All the money—all the value—is in the stock, and the new company is responsible for generating less total economic activity than there was before its existence. This is not an altogether bad thing—especially if we don't see economic activity as the goal—but it is an increasingly *true* thing.

Or think of it this way: A music industry that once supported everyone from musicians and writers to recording engineers, record manufacturers, distributors, illustrators, and store owners, is disrupted by MP3s and a few apps that let people play them. And those apps don't actually make money. Most often, they are money-losing propositions for everyone except the original investors, who have already executed their exit strategies. Amazon replaces thousands of brick-and-mortar stores, as well as all the industries that supported them—from window dressers to shelving manufacturers to the eateries where the shoppers lunched. Airbnb destroys far more jobs, income, and health insurance plans than it creates.

Instead of rethinking the innovations of the industrial age, we extended them into that "second machine age" envisioned by the MIT economists. Rather than transcending industrialism's antihuman values, we digitized them.

	ARTISANAL 1000–1300	INDUSTRIAL 1300–1990	DIGITAL INDUSTRIALISM 1990–2015
Direction	•	↗	⤴
Purpose	Subsistence	Growth	Exponential growth
Company	Family business	Chartered monopoly/ corporation	Platform monopoly (Amazon, Uber)
Currency	Market money (support trade)	Central currency (support banks)	Derivative instruments (leverage debt)
Investment	Direct investment	Stock markets	Algorithms
Production	Handmade (manuscript)	Mass-produced (printed book)	Replicable (file)
Marketing	Human face	Brand icon	Big data (prediction)
Communications	Personal contact	Mass media	Apps
Land & resources	Church commons	Colonization	Privatization
Wages	Paid for value (craftsperson)	Paid for time (employee)	Not paid/underpaid (independent contractor)
Scale	Local	National	Global
Optimized for	Creation of value	Extraction of value	Destruction of value

Some of these examples will be elucidated in the coming chapters, but what should already be clear is that the financial and marketing innovations we associate with the digital age are less disruptions than extensions of established business practices—new ways of exercising the same old corporatism. The chartered monopoly of the industrial age sees its expression in the platform monopolies of Uber and Amazon, and so on. Even unintended consequences, such as the havoc wrought by algorithms on the stock exchanges or of new technology companies on existing markets, find their origins in these embedded values of our industrial operating system.

Luckily, some companies are looking for another way out. Every executive who asks me how to overcome the growth requirement is actually another person in a position of power willing to look at what's going on and to try something new. CEOs are as likely candidates as any to supply human intervention here. After all, it's the CEOs who witness most directly how the free-market principles on which their businesses are based are themselves being undermined in the frenzy. Previous technological revolutions could be understood as creative destruction. It might have been bad for one's existing business, but at least there were new sectors where one could redeploy or invest. There's always a bull market somewhere. And to be fair, even in spite of the seventy-year trend toward lower corporate profitability over net worth,[38] even in spite of the fact that we've reached the limits of our planet's environment, there are still a few growth areas left. There are the morbidly robust sectors, such as military hardware and toxic cleanup. Right along with them are Big Pharma and medical insurance—industries that both grow along with the catastrophic health challenges wrought by war, pollution, and the stresses of a bad economy. Still, these are "crowded trades" already, and hardly safe refuge for the entirety of corporate America.

CEOs have tried hiding behind acquisitions, layoffs, outsourcing, and write-downs. Now, however, the smarter ones are looking to get in front of what appears to be an irreversible contraction. This is good news. Instead of waiting for government to intervene, they are choosing to intervene themselves. Besides, if this technological revolution is so inextricably married to the core values of the corporate program, it isn't going to spawn new opportunities for growth or labor, anyway. It is consuming the very landscape on which additional corporate activity would occur; at the same time, the highly centralized corporate model is being disrupted and in some cases replaced by the more distributed, peer-to-peer potentials of local and digital networks. Any way you look at it, traditional corporatism cannot survive indefinitely in a digital economy.

So here are some key concepts and strategies for moving forward. Each of them undermines at least one of corporatism's original core commands but ends up serving a corporation's constituencies better in the long run.

1. Get Over Growth

The only real choice for corporations that want to prosper in this volatile but ultimately contracting landscape is to reject the core commands of the chartered monopolies on which they are based, starting with growth itself. Instead of thinking of a company as an entity that must continue to show growth, think of it as an entity that must continue to generate enough revenue to pay its employees.

If infinite growth is no longer a possibility or even desirable, then you must shift your sights away from the big "win" in the form of an IPO, acquisition, or growth target. Instead, focus on achieving a more sustainable equilibrium. Think of it less like a war, where total victory is the only option, and a bit more like peace, where the objective is to find a way to keep it going. Running a business in this fashion feels less like a traditional football game with winners and losers and more like fantasy role-playing or a video arcade, where the object is to play as long as possible. Everything has to be kept viable, from the consumers and employees to the landscape in which you are operating.

Great family businesses have understood this for centuries: hire your friends and family, invest in people as if their personal fates matter to you, and think of your business less as a means of extraction than as a sustainable legacy. You can't "flip" your family, so why try to flip your business, your community, or your planet? Interestingly, while companies with more dispersed ownership structures tend to do a little better during boom economies, those run by families are a great deal more stable and profitable during downturns. This is because families tend to be less speculative with their own money and more risk averse when their children's futures are depending on the outcome of their decisions. As the chairman of Riso Gallo, an Italian rice producer since 1856, puts it, "We say that we didn't get the company from my parents, we are borrowing it from our children. And this is important. We are thinking of how it affects our offspring. We don't think in quarters, we think in generations."[39]

Unlike CEOs, who are incentivized to outperform each quarter, managers of family businesses are most concerned with increasing their odds of survival and finding good positions for their relatives. They carry little debt, make fewer and smaller acquisitions, retain their talent better, and enter foreign markets more patiently and organically. As a result, according to a *Harvard Business Review* study, the average long-term financial performance for family businesses exceeds that of their nonfamily peers by more than a few percentage points.[40]

Running a steady-state business means working against the extractive bias of the traditional corporation and looking instead toward investment and reinvestment in the markets on which the company depends. Customers, neighborhoods, and resources matter as much as the coffers. In fact, accumulating a war chest becomes understood as a liability—a dead zone of wasted capital that not only lowers profit over net worth but stands to lose value compared with inflation. Worse, the extraction of value from a company's own marketplace prevents that market from continuing to deliver dividends over time. It's like a loan shark killing his debtor; it may scare others, but it doesn't get the money back.

Instead, look to maximize ongoing revenue, stable profits, a healthy workforce, and a satisfied customer base. If anything, CEOs should be suspicious of sudden spikes in business activity and see them as potentially unsustainable growth trajectories. Rather than building a new factory to meet rapidly rising demand, the steady-state CEO rents temporary facilities to increase capacity while testing a new market's sustainability. Instead of using a temporary fad as an excuse to "grow the company," the CEO uses it as an exercise in resilience and scaling. The company should be equally ready to return to the previous size as to scale up even more—or take on some partners.

If a company can exist perpetually on a particular scale—be it a single shop, Web site, or factory—where is the obligation for it to continue to grow? For a business to find its appropriate size—even if this means scaling *down*—is not a Communist plot. Neither is creating a four-day workweek at full pay for the employees who got the company to sustainability in the first place. It may be counterintuitive in an age when CEOs are incentivized to grow now and pay later, but it is entirely more consonant with the efficiencies that digital technology brings to business.

The more Darwinian libertarians among us may feel compelled to reject a steady state as an affront to nature. True enough, nature is highly competitive, and species continue to adapt over time. On an evolutionary level, if you snooze, you lose. But this doesn't mean species have to keep growing, either in size or in population, to remain competitive. The ratio of owls to mice in a forest finds an optimal range, beneficial to both species. If the owls are too successful, they will run out of food. Likewise, an individual life form is allowed to become "full grown." Growth is part of an individual's childhood, not its adult phase. A world where everything is required to keep growing in order to stay alive makes no sense.

Instead, we must assemble businesses more in the fashion of ecosystems, such as a coral reef. Species can still iterate and adapt new approaches—evolve—but they are doing so within a greater stable matrix. Likewise, a steady-state business still makes progress. It still has research and development, but it is motivated by a need to serve better and more efficiently, not to fuel some artificially imposed growth target. The equivalence between growth and progress is not only artificial and unproductive but also unsustainable in a contracting marketplace and on a planet with limited resources.

In 2014, Toyota Motor Corporation publicly shrugged off a year of record growth and a $10 billion surge in operating profit because it saw overexpansion as a greater risk than slow or even negative growth. "If we make plans based on the pace of growth we have experienced over the last few years, things will not turn out well," one executive told Reuters.[41] Toyota's president, Akio Toyoda, went so far as to blame the company's aggressive expansion for both expensive product recalls (fittingly, cars accelerated by themselves) and its vulnerability to the economic crisis of the past decade. While competitor Volkswagen continues to pursue aggressive growth targets (ten million cars a year by 2018), Toyoda remains calm. "If a tree suddenly grows very fast," he explains, "the rings of the trunk will be unstable and the tree will be weak."[42]

2. Take a Hybrid Approach

Even CEOs willing to challenge the growth imperative can't turn their corporations on a dime without angering shareholders or violating their fiduciary responsibilities under their corporate constitutions. But they can begin testing new, sustainable strategies for a more distributed landscape by dedicating limited resources to them while still running their main businesses the old, extractive way. Such hybrid approaches give businesses the ability to feel more like they're merely hedging their bets on the future of the digital economy.

So, for example, one ingrained corporate behavior that might prove self-destructive in a digital climate is secrecy. Sometimes secrecy fails because of explicitly digital factors. For instance, proprietary security software for corporations and banks is proving less secure in the long term than open-source solutions. Why? Because secretly developed software is built and tested by fewer people in fewer scenarios. Open-source software has the benefit of hundreds or even thousands of developers, banging on it from all imaginable directions. Its openness is not a weakness but a strength. If a firewall's impenetrability is based on keeping how it works a secret, it's not a firewall at all; it's a security leak waiting to happen.

We can generalize on this principle to the greater business environment in a digital age. Today, open sharing and collaboration are proving better long-term corporate strategies than sequestering research and development. Hiding one's secret formulas suggests to the public—and to investors—that the company is depending on the innovations of the past and fears it won't continue to develop new ideas into the future. Its best days are behind it, and now all the company can do is play defense. In contrast, the confidently innovating company shares its developments in the hope of incorporating the insights of others. It welcomes contributions from the outside. People with great ideas can be hired. Companies that can identify areas for improvement can become new partners. If nothing else, demonstrating such openness expresses a company's expertise, leadership, and commitment to the culture it's supposed to be serving. Making better stuff is more important than who gets the credit.

But a company can't just open everything up at once, particularly when shareholders and others believe its best assets are its proprietary solutions. That's why some very established companies are developing quasi-open-source practices to connect internal R & D with outside technologists, universities, and, yes, even other corporations.

For instance, Procter & Gamble launched its "Connect+Develop" program[43] in 2004, with the goal of producing 50 percent of the company's total research and development output through collaborative innovation.[44] The C+D initiative presents outsiders with an easy path to partnership: P&G publicizes its innovation goals online and openly solicits solutions from anyone, large or small, who wants to collaborate. Connect+Develop maintains a "Needs" list on its Web site, where detailed project specs are visible to anyone with an Internet connection. Clearly, the company is hardly embarrassed by its openness to outside help; if anything, it treats its willingness to engage with outsiders as partners in problem solving as a demonstration of confidence.[45]

Over the last decade, the open approach has paid off. C+D's most famous win to date has been the Mr. Clean Magic Eraser, really just a block of melamine foam—a polymer invented by the German chemical company BASF. Originally branded as Basotect, the foam was intended for use as a soundproofer and, later, as insulation in automobiles. But when P&G technologists discovered that the foam also functioned as a cleaning sponge, they moved to partner with BASF. The two companies quickly packaged the foam as is, under the Mr. Clean Magic Eraser brand, and released it in 2003.[46] P&G and BASF then pooled their R & D teams to continue work on the product and released a wildly successful follow-up, Magic Eraser Duo, in 2004.[47]

Likewise, when P&G sought to create a long-term air freshener under its Febreze brand, it partnered with Zobele, an Italian firm with only five thousand employees and extensive, niche expertise in the peculiarities of manufacturing air fresheners. P&G had a winning concept for a slow-release air freshener that didn't require electricity. The company also had the supply and distribution chains to sell the innovation on a massive scale, but it lacked the expertise to engineer the idea into existence. Working with Zobele, P&G actually broke new ground and invented a no-energy automatic air freshener called Febreze Set & Refresh. The product went to market two years ahead of schedule, and the partnership continues to this day.[48]

In both cases, partner companies found new applications for their engineering while gaining access to the considerable branding and marketing power of Procter & Gamble. P&G, for its part, saved years of research and millions of dollars in development costs while annexing expert out-of-house think tanks in the chemical industry. Instead of acquiring outside companies in order to monopolize their research, P&G sought to partner with these companies and develop long-term relationships. Like a nation discovering the theory of comparative advantage, P&G did better by trading its branding and distribution expertise for these companies' particular skills. By letting them retain their independence, P&G also permits them to preserve their own innovative cultures—which can be called upon again in the future.

By edging into this strategy, Procter & Gamble avoids both risk and shareholder wrath. It is not completely abandoning its extractive and competitive corporate tactics all at once but merely "experimenting" with the more open-source development style of our decidedly more digital landscape. The company dedicates only a portion of its research and development efforts to agile, experimental processes while keeping the rest safely in the sphere of traditionally closed corporate practices.

Insulating legacy methodologies while simultaneously encouraging participation in newer approaches is called "dual transformation." As *Harvard Business Review* advises, innovation of institutional processes can take years to produce benefits. If an established company dedicates itself entirely to the emerging digital economy and its more open, peer-to-peer methodologies, "it throws away any advantage it still has." Its CEO also risks a shareholder revolt, or even civil action. To innovate safely, companies should conservatively reposition the core business while creating "a separate, disruptive business to develop the innovations that will become the source of future growth."[49]

Hybrid strategies can also give larger companies a way to contend with slowdowns or even a contraction in earnings. Instead of selling off assets, and doing so under pressure during a recession, companies can repurpose their unused facilities toward the very forces costing them business. This expands capabilities, develops competence in the new landscape, publicizes that competence, attracts new kinds of employees, and generates goodwill. Instead of simply resisting disruption, the hybrid company embraces it while also staking a real but limited claim on that disruption's becoming the new normal.

What might some of these strategies look like moving forward?

Consider supermarket chains, which are increasingly threatened by local food shares, community-supported agriculture (CSA), and growing discontent with Big Agra. The traditional corporate response to this conundrum is Whole Foods: a large, publicly traded company that attempts to provide consumers with organic products at scale. Problem is, "certified organic"—an appellation itself corrupted by corporate agriculture's lobbying of the Department of Agriculture—rarely means food from small or local farms. Six large California farms account for the vast majority of officially organic produce.[50] In spite of large posters throughout the stores featuring wholesome local farmers, locally grown food is still hard to find on the Whole Foods sales floor. That's because it doesn't make sense for a company of its size to sacrifice the efficiencies of scale to the minutiae of local sourcing and distribution. Whole Foods isn't a hybrid strategy at all but an industrial response to a new consumer trend.

Regular supermarkets might actually be in a better position to adopt a hybrid strategy toward local agriculture. Instead of seeing the growing CSA and farmers' market constituencies as competitors, they could embrace them as partners. If a supermarket were to encourage, or even provide the space for, a weekend farmers' market, it might make up for lost produce revenue with sales of condiments and dry goods—both of which sell at higher margins and with less waste than fresh goods. In terms of specialization, supermarkets are better at the distribution of long-distance packaged goods, while local farmers and CSAs are better at farm-to-table freshness.

Supermarkets are also better at cash management, electricity generation, parking, insurance, and a host of other things that vendors and consumers would be willing to pay for. There's no reason why farmers wouldn't pay for stall space, the same way they do at a municipal marketplace. Meanwhile, produce unsold by the end of the day could be stored in supermarket refrigerators or even sold wholesale to the supermarket for it to sell over the coming week. If local agriculture marketed in a more peer-to-peer fashion ends up replacing a significant portion of the produce and meat industry, such a supermarket will be positioned to prosper—without having surrendered its core business in the meantime.

Even a store such as Walmart can hedge against its own scorched-earth corporate policies by fulfilling the emerging need for more local, peer-to-peer marketplaces. Importing inexpensive goods from China and selling them to middle-class Americans is not a good long-term strategy, anyway—not when America's middle class is descending into poverty and the Chinese are becoming a middle class themselves. That's an arbitrage that has run its course, as the company's own numbers are indicating more clearly every quarter. If the resurrection of a middle class in America is going to depend largely on less extractive and more lateral economic activity, a company such as Walmart can ensure its own ongoing prosperity by finding a way to participate.

Instead of resisting the future with more aggressive tactics, Walmart must accept the impending contraction of its consumer base. This doesn't mean closing up shop or selling off real estate under pressure; it means repurposing at least some of its assets and expertise to the new environment. The company could try supporting the growth of peer-to-peer digital marketplaces such as Etsy by creating real-world analogues for them. The sales floor could become a hybrid of local crafts and boutiques alongside Walmart's own offerings, all funneled through its own checkout and inventory systems. Popular, locally produced items in one location could be quickly identified and then ordered for distribution to the most probable markets, using Walmart's vast trucking and fulfillment operations as well as its globally connected network of stores. Instead of simply extracting cash from communities, it would put itself in a position to foster local enterprise as well as peer-to-peer activity between communities that could not otherwise easily reach one another. The company could leverage a portion of its real estate, expertise, and logistics to enable and participate in the economy that may just replace it anyway.

When I shared this approach with researchers at the World Economic Forum, they argued that a company like Walmart cannot operate this way without violating its corporate charter and risking shareholder lawsuits. True enough; but while making other people wealthy may not be consistent with a traditional corporate code, it may just be a better long-term strategy for prosperity than economic destruction. In a shrinking, post-growth landscape, maybe keeping one's customers and suppliers viable is the best option. By facilitating local economic development, Walmart would be sustaining the markets on which it depends for its retail businesses. Or, in terms that shareholders would understand, the money will come back.

These are the sorts of principles that are already espoused, albeit a bit less dramatically, by a relatively new corporate ethos called "inclusive capitalism." Conceived by a task force including E. L. Rothschild CEO Lady Lynn Forester de Rothschild, the inclusive capitalism movement seeks to correct the dislocations caused by traditional corporate capitalism, particularly over the last thirty years. The initiative holds conferences and publishes papers encouraging corporations to conduct at least some portion of their business activities in ways designed to promote economic and social development.

One of its tenets is that larger corporations should attempt to award contracts and purchase supplies from as many small and local businesses as possible, in order to feed the local and bottom-up economy. Unlike cash transfers to fellow megaconglomerates, purchases from small and medium-sized businesses tend to put more money into circulation. The cash ends up injected into businesses that operate closer to the ground, where a higher portion of revenue goes to salaries and local taxes.

Corporations can do an awful lot of good for the world and provide insurance for themselves by adopting a hybrid approach to contract rewards. Hewlett-Packard UK, for example, contracted to over six hundred small and medium-sized enterprises (SMEs) in 2011, intentionally dedicating some 10 percent of its supply-chain budget to these smaller vendors. By doing so, HP ends up working against the perverse, winner-takes-all extremes of corporate efficiency in a digital age.[51] Moreover, utilizing the equal and opposite capability of digital technology to bring about more distributed outcomes, in March 2012, IBM helped to launch the Supplier Connection, an online marketplace connecting small business suppliers to large corporations looking to follow in HP's footsteps. Over $150 billion has moved through this network annually since its inception.[52]

It's the equivalent of a fisherman throwing back smaller fish to populate the lake on which he depends, or a farmer rotating a cash crop with one that restores the topsoil. Smart businesses know not to destroy the communities on which they are depending.

3. Change the Shareholder Mentality

Inspired by the inclusive capitalism initiative, the Unilever conglomerate, responsible for such brands as Lipton, Dove, and Hellmann's, has done away with its quarterly earnings reports in order to focus on advancing the social and sustainability interests that must be addressed if there is even to be a consumer market in the future.[53] Clearly, Unilever is still constitutionally obligated to deliver returns, but at least it's given itself a bit more room to engage in activities that might not immediately be recognizable as profitable.

Most corporations don't yet have the courage to follow suit—not even for initiatives that make pure business sense. According to a 2013 McKinsey survey, over half of corporate executives would pass on a viable project "if it would cause the company to even marginally miss its quarterly earnings target."[54] They are so afraid of their shareholders that they surrender what they believe to be in the best long-term interests of the company's profitability. These are not progressive, environmental, or spiritual priorities they're sacrificing to shareholder addiction to growth but *business* priorities.

In an economy in which corporate profit over net worth is declining, shareholders can no longer look to share price as the single metric of success. If a company can't grow without cannibalizing itself, then it has to give up on growth if it wants to survive in the long term. That means its stock just won't go up until the next period of growth, if that ever even happens. The good news is that it doesn't have to. A nongrowing company can be tremendously prosperous and deliver the vast majority of that prosperity to its shareholders. It just might not come in the form of a rising share price, which is the only metric most shareholders understand. That's why shareholders need to be trained to value other metrics or be replaced by people who do.

The easiest alternative to make shareholders happy is to pay them. Dividends—a quarterly payment to shareholders—used to be considered a good thing. An investor would buy a stock hoping for long-term growth but counting mostly on participating in the company's generation of wealth. In our current, growth-obsessed stock market, dividends are understood as a sign of weakness. Can't the company put that money to work? Companies that engage in particularly large regular payouts to shareholders are called "dividend traps," because investors fear a big dividend portends slower growth and a decline in share price.

We might better call stocks with no dividends "growth traps." If a company is depending entirely on quarter-over-quarter growth in order to deliver value to its shareholders, it is in a much more precarious position—particularly in a contracting economy—than a company that has managed to achieve sustainable prosperity. It's one thing to grow. It's another to be dependent on growth in order to pay back debts and generate shareholder value. Or, worse, to simply promise that real earnings are coming at some point in the future.

The disproportionate emphasis on share price is magnified further by our increasingly digital stock exchanges. Algorithms can trade only on changes in share price. They depend on volatility, not consistent returns. To an algorithm, a stable stock market is a profitless one. And a high-speed, high-frequency trading program never sticks around long enough to collect its dividend, anyway.

Instead of catering to this prosperity-defeating, ultrashort-term mentality, CEOs of sustainable companies need to communicate honestly with their shareholders about the firms' prospects—and the virtues of holding on to shares. Dividends and the stability of returns ought to replace share price as the measure of the company's value over time. Common shares become more like what are now known as the "preferred" shares usually owned by institutional investors and pension funds. Preferred shares don't shoot up so much when a stock is in favor but, instead, provide constant returns to investors. It's less like owning a share of stock than, well, owning a company.

Stressing earnings over share price also helps a company unwind from its own destructive impact on the greater economy. Unlike share prices, which remain trapped in capital, dividends come in the form of money. Sure, they can be reinvested in more shares, but they can also be spent. If a shareholder population can't be convinced to accept prosperity over growth, then they need to be replaced. This isn't so easy. The more people there are who want to sell their shares, the lower the shares' prices go, and that's when the class-action lawsuits start coming. The better alternative may be to buy out the shareholders oneself and get off the stock exchange altogether.

Going private—even temporarily—permits businesses to hit the reset button. They can eschew shareholder obsessions like growth targets and focus on the long term and on creating real value instead of the day's closing share price. Until recently, privatization has been used mostly as a form of financial jujitsu. A private-equity firm purchases a distressed or undervalued firm, uses its banking relationships to restructure the company's debt, and then puts it back on the market for a hefty profit. That's what happened when the Blackstone Group purchased the Hilton hotel chain in 2007, for $47.50 per share, a 32 percent premium to the stock's closing public valuation.[55] Seven years later, Blackstone took Hilton public once more, with an IPO that netted them a $10 billion profit—credited mostly to complicated tax and debt maneuvers.[56] Even at that, many observers considered it a disappointing return on investment.[57]

But buying back a company from one's shareholders can also give a CEO, a founder, or employees a chance to suspend growth and focus on more important things. That's the approach taken, perhaps most famously, by Michael Dell. While the company was growing in the 1980s and 1990s, shareholders were more than happy. But Dell found its growth slowing as the PC market was disrupted by tablets and the server market upended by cloud services. The company's problems were compounded by its typically corporate, growth-obsessed overexpansion, as well as its myopic focus on supply costs and margins over product innovation.[58] Michael Dell sought to implement some seemingly radical changes, such as selling PCs at a loss in order to upsell more profitable software and services, but his impatient shareholders disagreed.[59] Such a pivot would take longer than a single quarter to accomplish. If it failed, or failed to succeed fast enough, Dell's share price would decline further, making the company an easy target for a takeover, an ouster, or a shareholder lawsuit.

Using successful but contracting markets as loss leaders in order to pivot toward new opportunities is not a novel, radical approach at all. It is basic common sense, taught in any undergraduate business course. As Michael Dell explained in his SEC filing about the buyback, "These steps in Dell's transformation are needed to restore the Company to health in the long term. In the short term, however, they are likely to lower gross margins, raise the Company's operating expenses and raise capital expenditures, resulting in lower earnings."[60] Shareholders just won't put up with that. Moreover, share value had become such a widely accepted metric of corporate health that if the company's share price continued to drop, according to Michael Dell, it could hurt consumer perception and employee retention.[61] Unable to take the basic steps required to save a publicly held company, Dell enlisted the help of a private-equity firm to buy it back from his shareholders.

When the deal was done, he announced to his employees that Dell was the largest company in terms of revenue ever to go from public to private.[62] He did so, at least in part, to have the freedom to strengthen the company and its future instead of futilely trying to prop up its share price for stockholders who had no real interest in the company, its employees, or its sustainability.

Although Dell went private for purely business reasons, sometimes a company goes private to save what it sees as its soul. For example, CEO Mickey Drexler is largely credited with creating the brand magic that made the Gap synonymous with youth and effortless cool in the 1990s. Drexler accomplished this with an auteurlike sense of positioning, which he attempted to apply to the more high end J.Crew brand when he came on board there as CEO in 2003.[63] But after several years, Drexler found himself in combat with shareholders who expected the company to be working on margins and efficiencies—more like Zara or H&M, whose cash registers are networked to robotic supply chains capable of replacing stock in real time. Drexler convinced his board of directors to take the company private, largely to give himself the time and breathing room he needed to build the brand into something as iconic as the Gap.[64]

When a company does go private, it must select its private equity partners carefully. Otherwise, it is no guarantee against short-term pressure. As of this writing, Drexler and J.Crew are taking heat from their private equity stakeholders to go public again. It's been almost five years now, and the investors want to cash out, as well as fund a major international expansion. The growth imperative is hard to kill.[65]

Counterintuitively (at least to corporations that see their workers as expendable resources), the most patient shareholders might just be a company's own employees. Unlike retail shareholders, employees have a stake in the company's long-term prosperity. By becoming wholly or even partially employee owned, corporations can help ensure that business decisions reflect the best interests of the company, its employees, and their community—not distant investors or the abstract momentum of capital itself. It's still corporate capitalism, but corporate capitalism that ensures that the primary stockholders are also real-world stakeholders in multiple facets of the enterprise.

So far, most companies that have sold themselves to employees have done so under duress. For instance, in 1985, Amsted Industries, a heavy industrial parts manufacturer with some thirty-five plants across the U.S. and Canada, learned that corporate raider Charles Hurwitz had targeted the company for hostile takeover. That same year, Hurwitz would become famous for purchasing the Pacific Lumber Company and attempting to clear-cut the old-growth redwoods of California for a fast, onetime profit. So rather than accept its business being dismantled and sold off for its assets, Amsted opted to distribute ownership to its employees through a stock ownership plan, for a price of $529 million.[66]

Thirty years later, the company is still here and still employee owned. It has endured its share of bumps and conflicts along the way, and its shareholder communications sometimes sound more like a Brooklyn food co-op meeting than a traditional quarterly report. But the company's shift toward employee ownership attests to the possibility of similar arrangements, even for thriving companies looking to maintain prosperity instead of sacrficing it to growth.

Many companies set themselves up this way from the start, enjoy greater stability than their shareholder-owned peers, and still grow plenty big. With roughly 160,000 employees, Publix Super Markets is the seventh-largest privately held company in the United States, and the largest corporation whose employees own a majority stake.[67] Approximately 80 percent of the company stock is distributed among employees, while the rest belongs to the family of founder George W. Jenkins. Employees with a thousand hours and a year of employment receive an additional 8.5 percent of their pay in the form of stock. Stock price is set quarterly by an independent evaluator. According to *Forbes*, a store manager with twenty years at the company, in addition to a $100,000-plus salary, will likely own $300,000 in stock. That means that Publix employees, from the bag boy to the butcher, have no incentive to leave—especially since the company promotes almost exclusively from within.[68]

That kind of investment in developing and retaining individual employees doesn't show up on a ledger, but as Publix's ongoing success against Walmart grocery stores in its region proves, it does provide bottom-line value in the form of engaged, knowledgeable, and, yes, happy workers. Publix is the most profitable grocery-store chain nationwide, with a margin of 5.6 percent, compared to an industry average of 1.3 percent.[69]

4. Choose a New Operating System

Of course, Publix is at an advantage because it was built from the beginning as an employee-owned company. It is anything but public, and its employee shareholders are benefiting from the largesse of their founder. His company has ended up more prosperous as a result of his willingness to work against the more extractive biases of traditional corporations, but if he were the CEO of a publicly traded company, he'd likely be in court or worse.

As corporate law is currently structured, CEOs and their boards of directors can be held liable if they fail to do everything in their power to maximize quarterly returns for public shareholders. CEOs are not merely incentivized to pursue the short-term bottom line; they are legally obligated. Rather than liberating the corporation from such ultimately counterproductive rules, the digital age has put them on automatic, exacerbating their impact and making them appear more permanently embedded than ever. The more promising potential of the digital environment would be to revise the corporation itself to our liking. That's the invitation here—not to digitize the corporation with technology but to approach the corporation itself from a digital perspective of redesign. The corporation's charter can be recoded.

For example, many are toying with the "benefit corporation" as a way of tempering the emphasis on short-term and extractive profit suffered by traditional corporations goosed up on digital systems. A benefit corporation is expected to pursue profits, but that profit motive must be secondary to a stated social or environmental mission. By law, share price must take a backseat to something else, something decidedly beneficial. These corporations must develop metrics to measure social and environmental benefit based on third-party standards and then submit an annual report to government authorities confirming their compliance.[70]

Baby-food manufacturer Plum Organics is the largest certified B corp on the *Inc.* magazine list of the top 5,000 fastest-growing companies in America, coming in at number 253.[71] Launched in 2007, Plum did not register as a benefit corporation until it was about to be acquired by Campbell Soup Company in 2013. Plum had always committed itself to high environmental standards and appropriate treatment of its employees, and feared losing this leeway under the stricter supervision of a conglomerate's management and shareholders. By becoming a benefit corporation—the first to be acquired by a publicly traded company—Plum gave itself the insulation it needed from its new owner's shareholders.[72] Plum's stated mission is to produce organic baby food using lightweight packaging, pay its lowest-earning workers at least 50 percent above the living wage, and give away at least one million pouches of food to needy children per year.[73] As long as Plum remains a benefit corporation, that mission will be protected from any corporate or stockholder interference. Unlike its parent corporation, Plum is legally required to prioritize its goals of social and environmental good above that of profit.

Other companies have opted to become what are known as "flexible purpose" corporations, which allows them to emphasize pretty much any priority over profits—it doesn't even have to be explicitly beneficial to society at large.[74] Flexible purpose corporations also enjoy looser reporting standards than do benefit corporations.[75] Vicarious, a tech startup based in the Bay Area, is the sort of business for which the flex corp structure works well. Vicarious operates in the field of artificial intelligence and deep learning; its most celebrated project to date is an attempt to crack CAPTCHAs (those annoying tests of whether a user is human) using AI. Vicarious claims to have succeeded, and its first Turing test demonstrations appear to back up its claim.[76]

How would such a technology be deployed or monetized? Vicarious doesn't need to worry about that just yet. As a flexible purpose corporation, Vicarious can work with the long-term, big picture, experimental approach required to innovate in a still-emerging field such as AI. Although investors including Mark Zuckerberg and Peter Thiel have invested $56 million in the company, the flexible purpose structure prevents them from exerting the sort of pressure to get to market that venture capitalists typically put on their investments. The company can't be forced to sell out or to abandon scientific curiosity for commercial viability.

Vicarious has freed itself from the pressures of the market without having retreated to a research university, where funding comes with strings of its own. "[We] are not constrained by publication, grant applications, or product development cycles," one Vicarious statement reads. "At Vicarious, there is room to develop new approaches that would otherwise not be supported in academia or industry."[77] In effect, it's hacked the venture capital process. Vicarious's purpose is not social, per se, and not environmental either; it is the pursuit of knowledge. That purpose would not qualify the company for benefit corporation status. But by registering as a flexible purpose corporation, it can take on limitless investment while still placing exploration and experimentation above any demands for profitability that might arise.

Finally, the "low-profit limited liability company," or L3C, is a hybrid corporate structure first used in Vermont in 2008, tailor-made for the digital era's socially conscious entrepreneurs. It's a you-can-have-your-cake-and-eat-it-too approach to giving a company many of the benefits of a nonprofit charity while still offering its founders a way to raise venture capital and investors a way to cash out. An L3C works similarly to the flexible purpose corporation, in that it is not bound by the same rigorous reporting standards as the benefit corporation.[78] It can solicit funds from a greater range of potential investors, including private foundations, socially conscious for-profit entities, and grants, which each invest on a different tier suited to their goals and legal requirements. However, the L3C is limited to returning only modest profits.[79] That makes it a good structure for organizations such as Homeport New Orleans,[80] a small volunteer organization that cleans up marinas around Louisiana, or Battle-Bro,[81] a grassroots veteran-support network. It's a simpler, less burdensome structure than a nonprofit, suited for small groups operating with minimal revenue, for which the "win" is less about cashing out than about remaining financially sustainable. To the socially conscious investor, *some* profit is better than no profit. But is extracting profit really the best way to capture a company's value in a twenty-first-century digital economy?

Perhaps the real problem with the corporate program has less to do with social values than with simple math. Instead of attempting to mitigate the destructive power of a now digitally charged corporation on the world by inserting a socially beneficial purpose, we may better look at the underlying financial premise: The corporation was invented to extract circulating currency from the economy and transfer it into profit. No stated social benefit is likely to compensate for the social destruction caused by the corporate model itself. In other words, even if someone like Elon Musk or Richard Branson creates an earth-shatteringly beneficial new transportation or energy technology, the corporation he creates to make and market it may itself cause more harm than it repairs. Yes, such corporations bail some water out of the sinking ship, but they are, themselves, the cause of the leak.

In fact, none of these new corporate structures addresses the central flaw that precedes each of runaway capitalism's social, environmental, or economic excesses: the idea that more profit equates to more prosperity. Profit might lead to more shareholder value, but it doesn't necessarily maximize the wealth that could be generated by the enterprise over the long term and for everyone involved—even its founders.

That's why the not-for-profit, or NFP, might ultimately be the best model for the future of enterprise on a digital landscape.

Many mistake the term "nonprofit" (as the not-for-profit is also called) to mean "charity" or "volunteer." This isn't the case. The distinction lies in what is done with profits after expenses and salaries have been paid. Publicly traded corporations direct financial surpluses back to investors, CEOs, and boards of directors. They have little incentive to churn it back into the business and, as we have seen, a great deal of incentive to maximize surpluses at the expense of employees, the environment, and even the corporation itself.[82]

NFPs, on the other hand, are legally restricted from distributing surplus revenues to shareholders—stockholding investors, boards of directors, founders, worker-owners, or otherwise. An NFP must direct all surplus revenue back into the company. A not-for-profit's employees can be paid handsome salaries and may engage in a very broad range of work—broader even than benefit corporations. But an NFP may not have owners and may not be used to benefit anyone or anything other than the stated beneficiaries of its work.[83] It can make its employees wealthy and its customers happy, but no matter how successful it becomes, it can't extract cash out of the system, and it can't be sold.

The most creative digital companies have begun to mix and match various corporate strategies. The Mozilla Foundation, developers of the Web browser now called Firefox and one of the most successful digital companies of our time, is a not-for-profit. The company is a success both for its widely used open-source technologies and for its leadership position in a field dominated by platform monopolies. Mozilla is actually made up of two different entities: the Mozilla Foundation, a nonprofit, and the Mozilla Corporation, which the foundation oversees. The subsidiary corporation is responsible for much of Mozilla software's development, marketing, and distribution. It collects the massive revenue generated by Firefox,[84] but it has no publicly traded stock, no dividends, and no shareholders. All profits are directed back to the nonprofit,[85] which can spend them only to fulfill the Mozilla Foundation's nonprofit mission: "To promote the development of, public access to and adoption of the open source Mozilla web browsing and Internet application software."[86] By shunting profits into a nonprofit instead of delivering them to shareholders as capital gains, Mozilla is able to maintain its distributed network of open-source volunteers and five hundred to a thousand paid employees.[87] Capital is always in service of the business, its products, its employees, and its customers, never vice versa.

Any of these structural adjustments, and many others currently emerging, give corporations a way to transcend, or at least sidestep, the growth mandate that threatens their sustainability and longevity. Instead of removing money from the economy, they end up distributing their prosperity laterally—as if through a network. The money stays in circulation, providing currency to more people and enterprises. The piles of cash no longer accumulate, and the corporate obesity conundrum improves. Profit over net worth increases, even if only because net worth is shrinking. It's like getting a better body-fat percentage simply by losing weight.

But if corporations cease to grow or even get slimmer, if money is moving sideways between many people instead of upward from debtors to lenders and investors, then how does the entire economy grow? Isn't the whole system, from the Federal Reserve on down through the banks and bonds and GNP, all dependent on some rate of growth? Don't we always need more housing starts, more consumer activity, and increasing corporate earnings just to pay our national debt?

Yes, because that's the way the operating system was set up in the first place. Corporations may be the dominant players in the economic game, but that's only because they were built to succeed within the context of a very specific set of rules. If we want to see a genuinely new type of player emerge, we must rewrite the rules of the growth game itself.

Chapter Three

THE SPEED OF MONEY

COIN OF THE REALM

Imagine growing up in a world where only the Macintosh operating system existed. If every computer you ever saw ran OS X, you would never know there were any other possible systems. In fact, you wouldn't even know there was such a thing as an operating system. OS X would just be what a computer looks like. The same is true for money. The type of currency we use is the only one we know, so we assume its problems are part of the nature of money itself.

Something about our economy isn't working anymore. But it's hard to call attention to the flaws in our system or suggest improvements without challenging a few seemingly sacred truths about the money we use and what it was invented for. Since the Cold War—which is when we put "In God We Trust" on our money[1]—doing anything that called the tenets of capitalism into question was considered traitorous. We were steeped in an ideological war over a whole lot of things, but it seemed to be mostly about the freedom of corporate capitalism versus the tyranny of state Communism. To deconstruct anything about capital—where it came from, how it worked, what it favored—felt very much like deconstructing the American way. Inquiring into our currency's origins was to call attention to its very inventedness. And no good could come of that—not when the faith of investors, consumers, and borrowers was so inextricably tied to our prospects for growth.

Corporate grants and tenure appointments went to economists whose research confirmed the merits of capitalism; those whose work fell outside this purview were shunned or blacklisted.[2] As a result, the economists who passed through our finest universities ended up seeing the architecture of capitalism as a precondition for any marketplace. That's why the economic models to emerge over the last half century, however complex and intelligently conceived, almost invariably assume that the given circumstances of our particular market economy are fixed, preexisting conditions. Our economists have been trained in an intellectual universe with just one recognized economic rule set. That might be fine if economics were part of the natural world, whose principles are discovered through science, but it's not. The economy is less like a forest or a weather system than it is like a technology or a medium. It was created not by God but by people.

Indeed, if the chartered monopoly can be thought of as a piece of software, the central currency system on which it runs might best be understood as an operating system. The one we use—the bank-issued central currency of capitalism—is the only one most of us know. Even "foreign" money is just someone else's bank-issued central currency. Like the fictional computer users who know nothing but Macs, we think the stuff in our wallets or bank accounts is money, when it's really just one way of accomplishing some of money's functions.

Luckily, as members of a digital society, we are adopting more hands-on approaches to many of our most entrenched systems—or at least are expressing a willingness to understand them more programmatically. As we have seen, an individual corporation can be recoded, like a piece of software, to create and distribute more value to its various stakeholders. To do so, it must prioritize value creation and circulation over growth. The company may even stop growing or begin to shrink, which is perfectly okay as long as it spends down its frozen capital to satisfy its debtors and investors and then arrives at an appropriate scale for its market.

But while such companies may better serve the needs of people and even culture, they are incompatible with the underlying economic operating system. If corporations stop growing, then the economy stops growing, and unless something else comes in and changes the equation, the whole house of cards comes down. This is less a function of corporate greed or investor impatience than of the currency system we use and the fact that we use it to the exclusion of all others. Its universal acceptance has allowed currency to become a largely unrecognized player in the economy, as if it were an original feature of market activity, like supply, demand, labor, or commodities.

Currencies, tokens, and precious metals have indeed been used as means of exchange for thousands of years; but debt-based, interest-bearing, bank-issued central currency is a very particular tool with very particular biases—most significantly, a bias for growth. Capitalism itself is less the driver of this currency than it is the result. Capital is not an ideology so much as an artifact of a kind of money—a way of exploiting a particular operating system that runs on growth.

There used to be many kinds of money, all operating simultaneously. This may seem counterintuitive today, when the very point of money is to be able to count how much you have and compare it to how much everyone else has. But before the invention of central currency, money's primary purpose was to help people exchange goods with one another more efficiently than simple bartering allowed. Anything that promoted the circulation of goods between people was considered a plus.

In fact, prior to the emergence of the bazaar, most people didn't have any need for money, anyway. They were peasants and farmed the land of a noble in return for a bit of the crop for themselves. The only money was precious-metal coin, either left over from the Roman Empire or issued by one of the trading centers, such as Florence. A bit more currency was issued to pay for soldiers during the Crusades, and some of this returned home with the survivors, along with the crafts and technologies of foreign lands.[3]

As we saw, this gave rise to the bazaar, where locals traded crops and crafts with one another and purchased spices and other "imports" from the traveling merchants.[4] But there wasn't enough gold and silver coin in circulation for people to buy what they wanted. Precious metals were considered valuable in their own right. What little existed was hoarded, often by the already wealthy.

People bartered instead, but barter just wasn't capable of handling complex transactions. What if the shoemaker wants a chicken but the chicken farmer already has shoes? Barter facilitators arose to negotiate more complicated, multistep deals, much in the style of multiteam sports trades. So the shoes go to the oat miller, who gives oats to the wheelwright, who makes a wheel for the chicken farmer, who gives a chicken to the shoemaker. But the relative values of all these items were different, making the brokered barter system incapable of executing these complicated transactions with any efficiency.

That's when clever merchants invented currencies based on something other than precious metal. Instead, vendors whose sales over the course of a market day were fairly regular could issue paper receipts for the chickens or loaves of bread they knew they would sell by the end of the day: "This receipt is redeemable for one chicken at Mary's chicken stand." Market money could be issued by any merchant whose sales were stable enough to engender trust.[5]

So at the beginning of the market, the chicken farmer could spend her chicken receipts and the shoemaker could spend his shoe receipts on the items they needed, jump-starting the whole marketplace. The receipts would then circulate through the market throughout the day—just like money—until they got into the hands of people who needed chickens or shoes, at which point they would be exchanged with the original merchant for goods. To make matters even easier, the receipts would have a declared value stamped right on them—the market price of the products they represented. At the end of the day, extra receipts would be brought back to the merchants who issued them in return for metal coin, or saved for the next market day. The purpose of the money was not to make the issuer rich but to promote transactions in the marketplace and make everyone prosperous by getting trade moving.[6]

Almost anything could be represented as currency. Another very popular, longer-lasting form of money was the grain receipt. A farmer would bring his crop to the grain store and receive a written receipt for the amount of oats or barley he brought in. The receipt might be for a hundred pounds of grain, which had an equivalent numerical value in coin. It would usually be printed on thin metal foil, with perforations on it so that the farmer could tear off a piece and spend a portion at a time.[7]

Since the grain was already banked and in a facility that wasn't going anywhere, grain receipts tended to have a bit more long-term value. But they couldn't be hoarded like precious metals. Instead of gaining value over time, grain receipts lost value. The people running the storage facility had to be paid, and a certain amount of grain was lost to spoilage and vermin. So the issuing grain store reduced the value of the receipts by a specified amount each month or year. A hundred-pound receipt in March might be worth only ninety pounds of grain by December.

Again, this wasn't so much a problem as a feature of this money. People were incentivized to spend receipts as soon as they received them. Money moved through the economy quickly, encouraging transactions. Ideally, someone who needed grain ended up with the receipt just before its next date of reduction and redeemed it for the oats.

These local moneys worked right alongside the long-distance precious-metals currencies. Gold coins and silver pennies were still required by traveling merchants, who had no real use for stored grain or a future pair of shoes. They also provided easy metrics through which to denominate all those local currencies. The declared value of a loaf of bread on a bread receipt could be some fraction of a gold coin, making it easier for consumers to negotiate transactions, as well as for issuers to reconcile unredeemed receipts with one another at the end of the day.

The lords and monarchs tolerated all this trade for a while but began to resent people putting real monetary values on their self-issued currencies. Besides, the more people traded laterally, that is, with one another, the less dependent they were on the aristocracy. The peasants were getting wealthy from the bottom up, in an economy whose strength was based on the robustness of its transactions. Growth, a happy side effect of their increased capacity to transact, had to be appropriated. In doing so, it was turned into a financial weapon.

The nobles hired financial advisors—mostly Moors, who had more advanced arithmetic techniques than the financiers of Europe—to come up with monetary innovations through which the wealthy could retain their class advantages over the rising middle class. We already saw how the chartered monopoly would give royals the ability to assign entire industries to particular companies in return for stock in the enterprise.[8] But not every industry was that scalable—at least not back then. Kings also needed a way to extract value from all those little transactions between people and, ideally, to slow down all that economic activity so that the middle class did not overtake them.

So one by one, the monarchs of the late Middle Ages and early Renaissance outlawed local currencies and replaced them with what amounted to coin of the realm.[9] By law, people were forbidden to use any other currency—a rule officially justified, ironically, by the fact that the non-Christian icons of the Muslims appeared on some of the coins people had been using since the Crusades.[10] The real reason, of course, is that with absolute control over coin, monarchs could exert absolute control over their economies. People protested and much blood was shed, but they lost the right to issue their own currencies. Instead, all money would be coined by the king's treasury. As many economic historians have noted, this allowed the monarch to tax the people simply by debasing the currency and keeping the extra gold. What these same historians seem loath to point out, however, is that monarchs made money simply by issuing coin.[11] The monetary system itself gave those who owned capital a way to grow it.

In a practice analogous to the way central banks issue currency to this day, monarchs created coin by *lending* it into existence. If a merchant wanted cash to purchase supplies or inventory, he needed to borrow it from the king's treasury, then pay it back with interest. It was a bet on future growth. Unlike market money, which had no fees, or grain receipts, whose fees went toward a working grain store, central currency cost money. If people wanted to use money, they would have to pay for the privilege.

This was a brilliant, if exploitative, innovation: money whose core function was to make wealthy people more wealthy. Since the aristocrats already had wealth, they were the only ones who could participate in the new supply side of money lending. If people and businesses in the real economy wanted to purchase anything, they would have to get some cash from the central treasury. Then they'd have to use that money to make some deals and somehow end up with more money than they started with. Otherwise, there was no way to pay the lender back the principal and the additional interest.

So if a merchant borrowed a thousand coins from the treasury or its local agents, he might have to pay back twelve hundred by the end of the year. Where did the other two hundred come from? Either from someone else who went bankrupt (and was therefore facing debtors' prison) or, in the best case, from some new borrower. As long as there was more new business, there was more money being borrowed to pay the interest on the money borrowed earlier. This was great for the wealthy, who could sit back and earn money simply by having money.

Participants in the bazaar didn't fare so well. This new money was still scarce and expensive. Where market money was as plentiful as the demand for goods at the market, central currency was only as abundant as the participants' credit. Merchants who used to cooperate now competed against one another for coin in order to have enough to pay back their loans. Frequent currency debasements also led people to hoard money out of fear that current coin had more gold in it than whatever was coming next. Moreover, everything in the market now cost more, because money itself was extracting value from people in the form of interest. They weren't just paying for a chicken, but also for the chicken farmer's debt overhead. If they were participating in growth businesses, they might have stood a chance of keeping up with the cost of capital. But these were largely subsistence enterprises.

In country after country where local moneys were abolished in favor of interest-bearing central currency, people fell into poverty, health declined, and society deteriorated[12] by all measures. Even the plague can be traced to the collapse of the marketplace of the late Middle Ages and the shift toward extractive currencies and urban wage labor.

The new scheme instead favored bigger players, such as chartered monopolies, which had better access to capital than regular little businesses and more means of paying back the interest. When monarchs and their favored merchants founded the first corporations, the idea that they would be obligated to grow didn't look like such a problem. They had their nations' governments and armies on their side—usually as direct investors in their projects. For the Dutch East India Company to grow was as simple as sending a few warships to a new region of the world, taking the land, and enslaving its people.

If this sounds a bit like the borrowing advantages enjoyed today by companies like Walmart and Amazon, that's because it's essentially the same money system in operation, favoring the same sorts of players. Yet however powerful the favored corporations may appear, they are really just the engines through which the larger money system extracts value from everyone's economic activity. Even megacorporations are like competing apps on a universally accepted, barely acknowledged smartphone operating system. Their own survival is utterly dependent on their ability to grow capital for their debtors and investors.

Central currency is the transactional tool that has overwhelmed business itself; money is the tail wagging the economy's dog. Financial services, slowly but inevitably, become the biggest players in the economy. Between the 1950s and 2006, the percentage of the economy (as measured by GDP) represented by the financial sector more than doubled, from 3 percent to 7.5 percent.[13] This is why, as Thomas Piketty demonstrated in *Capital in the Twenty-First Century*, the rate of return on capital exceeds the growth rate of the economy.[14] Money makes money faster than people or companies can create value. The richest people and companies should, therefore, position themselves as far away from working or creating things, and as close to the money spigot, as possible.

Some companies, such as General Electric in the 1980s, understood this principle quite well and acted on it. They came to realize that their core enterprises were really just in service of the much more profitable banking industry. GE's CEO Jack Welch determined that the company made less money making and selling washing machines than it did lending money to consumers to purchase those washing machines. So he began selling off GE's factory businesses and turning the corporation into more of a financial services company. The washing-machine companies were sold to the Chinese. The new GE made loans, sold insurance, and provided capital leasing.

This worked quite well for the company and for those who followed in Jack Welch's footsteps.[15] His approach was canonized by Harvard and other top business schools, which began training their graduates to see productive industries as mere stepping-stones to becoming holding companies. The further up the money chain you can get—the more like a bank issuing money—the better.

The American economy became almost entirely dependent on its companies' and citizens' willingness to use credit. It didn't matter what they bought with that credit. Most of it went to Chinese goods, but that didn't matter. We were growing not the real economy of goods and services but the synthetic economy of money itself. The Western companies at the top of the food chain were selling credit, not consumables. The perfect productless product. Those unwilling to participate were researched by psychologists, then subjected to techniques of "behavioral finance" until they got with the program.[16]

To many of us, the whole system seemed to be working. Regular people began taking mortgages out on their homes every time real estate values went up, as a way of generating more capital for themselves. Even Alan Greenspan thought the triumph of capital and credit meant we had embarked on a new era of riskless investing. As the former chairman of the Federal Reserve remarked in 2000: "I believe that the general growth in large [financial] institutions has occurred in the context of an underlying structure of markets in which many of the larger risks are dramatically—I should say, fully—hedged."[17] That is, right up until the crash of 2007, when there turned out not to be enough real economic activity to support the overcrowded field of moneylending.

Like other market crashes, this one was blamed on dodgy deals and abuse of the system. Unwilling or incapable of looking critically at money, chroniclers of capitalism's excesses focus instead on human greed. The evil CEOs of Enron or WorldCom, the heartless bankers of Wall Street, or the overzealous traders at Goldman Sachs are indicted for fleecing shareholders and undermining an otherwise functional system. Parading them in handcuffs before the press makes us feel good for a moment, until we take in the truly clueless expressions on their faces. They thought they were just doing their jobs—trying to grow their companies at the rate dictated by their debt structures. Their behavior is normative, not an aberration.

Not that these characters are blameless, but when we fault "corruption" for our economic woes, we are implying that something initially pure has been corrupted by some bad actors—like a digital file that was once intact but whose data now has errors in it. That is not the case here. Rather, an economic operating system designed by thirteenth-century Moorish accountants looking for a way to preserve the aristocracy of Europe has worked as promised. It turned the marketplace into one giant debtors' prison. It is not only unfit for the needs of a twenty-first-century digital society; central currency is the core mechanism of the growth trap.

This is the real cause of the severity and longevity of the 2007 crash. Rather than figuring out how to compensate for central currency's extractive bias, a highly digital finance industry chose to exploit it. The digital perspective that allows us to see money as an operating system doesn't necessarily motivate people to revise the core code so that it serves people better. That would be a pretty heavy lift, even for the most idealistic among us. So instead, bankers and financiers sought to leverage the structural flaws of the money system for their own gain.

They understood that return on capital outpaces real growth and that in a digitally accelerated marketplace a disconnect between winners and losers was inevitable—just as on iTunes and Amazon. Only here, the winners would be those who had capital, and the losers would be those stuck in the real economy of goods and services. So they sold money to borrowers, then sold those loans to less-intelligent lenders. Meanwhile, they insured *and* bet against those original loans. This created a win-win for those with capital—they got paid their regular interest rates for lending money, but they also won their bets against the people and companies they were counting on to fail.[18] The lucky winners benefited not only from the return on capital but also from the inability of real growth to keep up with interest.

Instead of our cynically profiting from the failure of people and business to keep up with the needs of capital, might there be a way to change the way capital functions? Might money have gotten too expensive for its own and our good? We've seen how individual companies can wind down from the growth imperative by lightening their debt loads, changing investor expectations, and becoming more adaptable to market conditions on the ground. But the economic operating system on which these corporations are required to function cannot enjoy the luxury of voluntary slowdown. Unlike natural systems or even human society, an interest-based economy must grow in order to survive. This worked for centuries, as long as there were new regions to conquer, resources to extract, and people to exploit. An expansionist economic system both necessitated and inspired the colonizing of the Americas, Africa, and Asia. As long as more money was being borrowed, there was more money to pay back the bank and keep the currency afloat.

In the absence of new continents in which to expand growth, industry strived to speed up rates of production or to make existing processes more capital intensive. Industrial farming, for example, generates more crops in the short term than do traditional, less-intensive methods. It also requires more machinery, fertilizer, and chemicals. By abandoning the practice of rotating crops, industrial farming also depletes topsoil faster, which in turn generates even more dependency on chemicals and pesticides. More money is required—and that's the object of the game. If Big

Agra processes lead to a less-healthy population or higher cancer rates, Big Pharma is ready with costly fixes, fueling another source of economic expansion.

But we're all getting depleted in the process. Our real world of humans, soil, and aquifers replenish themselves more slowly than the impatience of capital can accommodate. "Housing starts" can accelerate only as fast as the market for new homes. When the marketplace isn't being artificially goosed by speculators, humans just can't keep up with the housing industry's need for excuses to cut down more forests, irrigate more land, and construct more homes. Moreover, as Naomi Klein has more than demonstrated in her book *This Changes Everything*, climate change is a direct result of an expansionist economy: the physical environment can't service the pace of capital while also sustaining human life.[19]

Economic philosopher John Stuart Mill identified this problem as far back as the 1800s. "The increase of wealth is not boundless," he wrote.[20] He believed that growth wasn't a permanent feature of the economy because nothing can grow forever. No matter what the balance sheet may be asking for, economic growth is limited by the finiteness of the real world. We can generate only so much activity and extract only so many resources. Instead, Mill saw the end of growth concluding in what he called the "stationary state"—a sustainable equilibrium in which there would be "a well-paid and affluent body of labourers; no enormous fortunes, except what were earned and accumulated during a single lifetime; but a much larger body of persons than at present, not only exempt from the coarser toils, but with sufficient leisure, both physical and mental, from mechanical details, to cultivate freely the graces of life."[21]

Mill didn't see this stationary state as devoid of improvements to society, technology, and overall satisfaction. It would merely mark the happy end of the era of big investment and associated extraction and growth. Capital will have fulfilled its purpose in building out our society and bringing us to the "carrying capacity" of our planet. In other words, people might produce and consume different or better stuff, but they won't be able to produce or consume *more* stuff. They won't have to, though, because the "progressive economics" of capitalism will have been abandoned and replaced with something else.

Digital technology, computers, networks, and miniaturization at first appeared to herald a shift toward these more steady-state approaches to life. Telecommuting would mean less gas consumption, and computer screens would mean less printing. Neither prediction came true. Worse, though, speculators saw in digital technology a gateway to a new, virtual form of colonialism: a new place to lend and deploy capital, new territory for growth.

Alas, the big data profiles of teenagers can't support the same robustness of growth as entire continents of slaves and spices. Besides, consumer research is all about winning some portion of a fixed number of purchases. It doesn't create more consumption. If anything, technological solutions tend to make markets smaller and less likely to spawn associated industries in shipping, resource management, and labor services. They make the differential between real growth and return on capital worse, not better. This means they push the banks and investors even further away from anything like real earnings until eventually there's a complete disconnect between capital and value.

So how can an entire economy that is based on an arithmetic premise of perpetual, infinite, and impossible growth somehow deleverage itself? And even if there's room for more growth, how do we unpack some of the accumulated frozen capital and get it back into circulation so people can buy and sell more goods and services?

Policymakers have painfully few good stimulative tools at their disposal, and the ones they do have aren't particularly good at getting money into the hands of real people, where it's needed. Helping people transact isn't the sort of activity their operating system was built to support in the first place. Central currency was intended to extract value from the economy, not pump it in. That's why it's more intuitive and superficially consistent to demand austerity and belt-tightening from debtor nations than it is to ease policy or lower interest rates. If countries or regions can't pay back the interest on the currency they "borrow," then shouldn't they be given less of it? That's the justification against bailing out failing eurozone nations. It sounds logical on the surface, but if the struggling nations are loaned less, then how are they to pay back more—especially if they're not allowed to print their own money to foster some economic activity?

Central banks weren't built to fix this. The Federal Reserve's primary function is to protect the wealthy—those who are holding cash—by preventing the inflation that would make that cash less valuable. During hard times, a compassionate central bank can choose instead to pump more money into the economy—but it really has only two ways to accomplish that. It can lend money to banks at the lowest interest rate possible—even zero—or it can buy the banks' stashes of bonds (what's known as "quantitative easing"). But for this money to reach the real economy, the banks still have to lend it to people and businesses. Nothing is forcing them to do that part, and in a low-interest environment, their profit margins on lending are squeezed anyway. Most banks would rather invest the money in more leveraged financial instruments or buy the stock of existing companies. Moreover, given the slow-growth economy, many banks refuse to take money from the Fed, loath to take on credit that they know they'll have to pay back. They already have more money sitting around than they know what to do with, and if they take on additional capital, their ratio of profit over net worth only gets worse.

The other choice is for the government to take on debt by borrowing money from the central bank and giving it to workers—ideally, people who are doing some task for the government, such as building infrastructure or providing a social service. The money these workers earn as payroll is then circulated through the rest of the economy when they make purchases. When the economy recovers, the government collects more taxes to pay back the central bank.

That strategy, employed successfully during the Great Depression, would be a tough sell today. Many of our elected leaders don't understand concepts as simple as the debt ceiling; most Americans don't realize that federal spending has been flat or down since 2009, lower than at any time during the Reagan administration, and even lower than Paul Ryan's infamous budget proposal of 2011.[22] They don't even see how improving infrastructure can itself stimulate economic activity[23] or that the very best time for the government to borrow money for that purpose is when interest rates are close to zero. It still feels like charity or socialism.

What people have a hard time wrapping their heads around is that putting money into circulation should be less about paying people for working or not working than it is about giving people a means to transact. We don't need the government to hire unnecessary workers; we need people to be able to exchange value with one another. Cash could serve that utilitarian purpose if it weren't so wrapped up in its other, more extractive function. This is the heart of the divide between the supposed 1 percent and the people. Our ability to generate value has been paralyzed by our inability to find a means of exchange. People who work for a living are suffering under a system designed to favor those who make their money with money. Yes, it's what Marx was saying; but there's an out. We're not looking at a fundamental property of economic activity or even an unintended consequence of capitalism. This is an economic operating system working as it was programmed to. And we can program it differently.

REPROGRAMMING MONEY—BANK VAULTS TO BLOCKCHAINS

Thankfully, we have both the perspective and the tools required to change the operating system of money, either by adjusting the one we use or by building some new ones. Although business intransigence and government incompetence will likely forestall any meaningful modifications to the central currency system, the greater digital landscape fosters alternative approaches to enabling transaction.

Besides, there's no reason to ask central currency to do something other than what it was programmed to do. It's a great tool for storing and growing wealth, for long-distance trading, and for large-scale, expansionist investing. It's just not a great tool for transactions between smaller players or for keeping money in live circulation. So let's not use it for that. Just as we don't ask a carpenter to build a house using a hammer but no saw, we can't expect the economy to function with just one monetary tool. Contrary to our intuition, we can have more than one form of money in operation at the same time. This wouldn't be a Communist plot at all; on the contrary, we would merely be subjecting currency to the open competition of a free market. May the best money or moneys win.

If we're going to consider remaking money for a digital age, however, we have to decide just what we want it to do. In programmer-speak, what are we programming *for*? The various answers to this very simple question lay bare the biases underlying many of the loudest proposals for changes to our currency system. For instance, what does pushing for a gold standard accomplish other than raising the portfolios of those who have already invested in gold? Requiring the Treasury to back every dollar with a certain amount of gold would certainly prevent the central bank from going on a printing spree. Money would get a fixed value, and there would be no threat of inflation. But how would a gold standard promote circulation over hoarding? It wouldn't. A gold standard is optimized to address fear that one's savings are not safe if they're measured in government-backed dollars. But gold-backed currency would be no better at promoting a peer-to-peer marketplace than gold coins were back in the Middle Ages. It's biased toward scarcity.

Bernard Lietaer, one of the economists who helped design the euro, has been proposing since 1991 that fiat currencies—money declared legal by the government but not backed by a physical commodity—be replaced or at least augmented with currencies that represent a "basket" of commodities.[24] His current suggestion is to create a currency that is backed by one third gold, one third forests, and one third highways. The gold is the fixed-commodity component, as there is only so much of it. Forests are the growth component; trees grow. And highways, thanks to tolls, are the income component. As an investor's response to deflation, or even as a new reserve currency, it makes sense.

But if we're trying to compensate for the way central currency tends to work its way out of circulation and into the bank accounts of the already wealthy, we should be looking instead for ways to help money move around better. This has less to do with making sure money has some intrinsic value for long-term storage and accumulation into the future, and a lot more to do with making sure it can serve as a medium for exchange right now. In economic programming terms, we should optimize no longer for the growth of money but for the *velocity* of money. Not for saving money but for exchanging it. By analogy to another newly digitized medium, it's less about finding a way to preserve movies, such as videotapes or DVDs, than about finding a way to distribute them to people's homes, such as through digital cable and satellite. We're less concerned with the content itself than with promoting its movement. In that respect, the money itself doesn't matter, anyway, except insofar as it helps people exchange goods and services. Perhaps that's why most of the first real innovations in digital currency had to do less with new kinds of money than with new means of transferring it.

For instance, the first online selling platforms, most notably eBay, turned millions of regular people into vendors for the first time. But there was no easy way for them to accept payments. Checks could take weeks to clear, stalling delivery unless sellers were willing to ship without funds verification. Credit cards were impractical: most casual sellers didn't make enough sales to offset the costs of a business account and payment processing system.

PayPal created the first utility capable of addressing the rising need for peer-to-peer transactions. The original model was simple. Buyers and sellers registered their bank accounts or credit cards with PayPal. The buyers authorized electronic transfers to PayPal. PayPal then informed the seller that the funds were secure, and the seller mailed the merchandise. The buyer verified its arrival, and PayPal released the funds to the seller. PayPal served as both a trusted exchange agent and an escrow account. The whole service was free, since PayPal could earn interest on the money during the three or four days it held it in escrow.

But the banking industry and its regulators sensed an upstart in the making and challenged the company's legality. Only regulated savings institutions are entitled to make money on "the float," as PayPal was doing. So PayPal changed its business model and began charging buyers and sellers directly for the service.[25]

Still, PayPal was the first of many companies to promote peer-to-peer transactions by lowering the barriers to entry into the existing money networks. A company called Square took this a step further, developing both the technology and the accounting infrastructure through which people could swipe credit cards and accept payments through their smartphones. Although many coffee shops and smaller retail stores now use Square and an iPad in place of more cumbersome and expensive credit-card systems, the people most dramatically empowered by the system were independent sellers and service people and those who want to pay them. Google and Apple, meanwhile, are competing to develop new ways of using credit cards and bank accounts through phones and tablets—technologies that, presumably, could help the smallest businesses as much as the biggest ones.

These systems increase the velocity of money by fostering transactions between nontraditional players, making them simpler to execute, easier to verify, and faster to complete. So far, however, they all use the same, rather expensive transaction networks—such as those run by the credit-card companies or the automated clearinghouse (ACH) system that serves banks.[26] In fact, they're really just digital dashboards for the existing trust authorities, which still validate every transaction and absorb part of the cost of fraudulent transfers—currently over $10 billion per year.

Don't cry for them yet; this is merely another way for them to justify their transaction fees. Without bad players, remember, the trusted authorities wouldn't be necessary. Credit-card companies are earning 3 or 4 percent on every purchase. That's more than the growth rate of the entire economy. And it doesn't even account for the primary source of credit-card company revenue, which is all the interest customers are paying (or further accumulating) on their balances. When a whole marketplace is not only paying up to the bank in the form of debt-based money but also paying a trusted authority to verify transactions, marginal costs become unsustainable. Merchants must mark up their prices to account for all the transaction fees, and commerce slows. Only giant retailers, with the ability to borrow money less expensively or even offer their own credit cards, are capable of reducing these fixed costs by filling some of the roles of the trusted central authority themselves.

Besides, as the ever-increasing frequency of major credit-card and consumer-information theft has shown, none of these systems is particularly safe. The trusted central authorities really aren't so good at what they do. From Target to J.P. Morgan, a single cash register or employee laptop can become the entry point for a systemwide hack, rendering all users vulnerable to credit fraud and identity theft. These companies' security services are fast becoming loss leaders for their more lucrative lending schemes.

Viewed in this light, this first generation of digital transaction networks are not revolutionary but reactionary. They ensure that the newly decentralized marketplace remains entirely dependent on the same centralized institutions to conduct any business. Meanwhile, the currency they employ—bank-issued reserve notes—is itself the product of a trusted central authority that also charges for its services. This, as we have seen, is an even bigger drag on the potential velocity of money.

What good is a distributed network like the Internet if all the actors on it still depend on central authorities in order to engage in peer-to-peer activity? How is it truly peer-to-peer if it goes through a central clearinghouse? It's still a bunch of decentralized individuals, each interacting with a monopoly platform—a new front end on the same old system.

These are the problems that the next generation of digital transaction networks are aiming to address. How can a distributed network of participants transfer and verify value collectively, without the need for a central authority? Is that even possible? Could a money system look and act less like iTunes and more like BitTorrent, where, instead of depending on a platform monopoly to negotiate everything, all the participants use protocols to interact with one another directly? Could a digital money system achieve with openness what traditional banks do with secrecy?

The only way to find out is to start as openly as possible. That's why Bitcoin first appeared as the subject of a 2008 white paper authored by someone (or multiple someones) going under the name Satoshi Nakamoto. The paper outlined a concept for a virtual currency created and traded on a peer-to-peer, open-source platform. It would need no central authority to issue it, nor any central middleman to verify or administer its transactions. The network platform would be called Bitcoin, and its currency would be called bitcoins.[27]

This idea was not entirely new. Virtual and decentralized currencies had been tried in the past. But what set Bitcoin apart was its proposed method for ensuring the legitimacy of these transactions. As Nakamoto explained, in order for a currency to function as a medium of exchange, it must meet two basic standards: First, users must be reasonably certain that the currency they hold is not counterfeit. Second, the currency can't be "double-spent"—that is, an unscrupulous buyer can't spend the same money on two separate transactions. Meeting these standards is fairly simple for centralized currencies. High-tech printing techniques discourage counterfeiting. Credit and banking clearinghouses offer protection against double spending by maintaining secure ledgers of people's accounts; spend money, and it is immediately subtracted from the single, centralized ledger.[28]

Nakamoto's paper proposed that a distributed network could generate even greater security than a centralized money system if users pooled their computing resources to maintain a collective and open ledger of their own. He outlined the proposed technology, thousands of people commented and made suggestions, and in 2009 the Bitcoin network was launched.[29]

Understanding how Bitcoin works isn't crucial to being able to use it, any more than understanding the chain of possession of electronic ballots is crucial to being able to cast a vote. But the more we understand Bitcoin's technology, the more we can trust it without relying solely on the word of those more digitally literate than ourselves. That's why Bitcoin's code is published and open source: if you're afraid there's some government or criminal in there running things, just look at the code and you'll see what's going on. I'll explain it here briefly, but the main takeaway is that there's no one in charge—which means the biases of Bitcoin are very different from those of a traditional interest-generating money system. This is a money system that works through protocols—digital handshakes between peers—instead of establishing security through central authorities.

Bitcoin is based on a database known as the "blockchain." The blockchain is a public ledger of every bitcoin transaction ever. It doesn't sit on a server at a bank or in the basement of a credit-card company's headquarters; it lives on the computers of everyone in the Bitcoin network. When bitcoins are transacted, an algorithm corresponding to that transaction is "published" to the blockchain. The algorithm is just a description of the transaction itself, as in "2 bitcoins came from A and went to B." Instead of a list of users and their bitcoin balances, the ledger simply lists the transactions in chronological order. It doesn't follow people, it follows the money. It's not a record of how much you have as much as a record of where the money came from and where it is going to.[30, 31]

To get a transaction into the ledger, two users must first agree to the exchange. Using a pretty standard form of cryptography (public and private "keys"), both users "sign" the intended transaction, at which point it is broadcast to the network. Immediately, other members of the network who have devoted some of their computers' power to the Bitcoin process begin verifying the transaction and committing it to the public ledger. This part involves solving a bunch of computational puzzles—a way of guaranteeing that a whole lot of different computers have verified the transaction before it goes in. This prevents one bad actor from posting fake transactions into the ledger. He'd need more computing power himself than the whole network of thousands of users in order to overpower them. When enough people verify the transaction, it becomes part of the permanent ledger—part of a new block of transactions, recorded in the chain. (That takes about ten minutes, compared with a bank, which might take up to a week to confirm new funds.) As the system gets up to speed, people who verify and maintain the blockchain are rewarded with bitcoins. That process is called "mining," and it is how new bitcoins enter circulation. This dilution of the money supply, as such, is infinitesimal compared with credit-card fees, and its drag on transactions is negligible.[32, 33]

Nakamoto, it seemed, had at last developed a way to distribute trust in the digital economy: create a public record of transactions and lock it down, not with bank security or virtual firewalls, but with the combined computing power of the community. Your money can't be stolen, because there's nowhere to break into. Everyone has a record of everything.

For almost five years, the Bitcoin network and its pool of bitcoins grew, while users exchanged bitcoins for products such as thumb drives, alpaca socks, and, yes, drugs. The fact that people transact through cryptographic keys instead of names or e-mail addresses lets them make purchases anonymously. Since there's no credit-card statement at the end of the month listing the illicit goods and services someone may have purchased, cryptocurrency became popular on black markets and earned a reputation as money for criminals. Then in late 2013, something interesting, if all too predictable, happened. Whether in response to the high-profile bust of an illicit online bitcoin-based marketplace known as the Silk Road or to the growing participation of Chinese users, Wall Street suddenly seized on bitcoins as a new instrument for speculative investing. It became the next big thing.[34]

What speculators love so much about bitcoins is that only a limited number of them will be mined into circulation. The mining process will be complete within the next couple of decades, and the money supply will remain fixed after that. Moreover, if a user loses his or her private key, then that user's bitcoins leave circulation. These two factors make the bitcoin currency highly deflationary. If they were actually to catch on, investors reasoned, those who got in early would have cornered the market on an entire currency. That's why from 2012 to 2013, the price of a single bitcoin skyrocketed, from ten dollars in November 2012 to a thousand dollars a year later.[35] There are now bitcoin investment funds—one famously started by the Winklevoss twins, known best for hiring college student Mark Zuckerberg to build their social network platform and subsequently losing it to him.

They may be missing the nature of this opportunity as well.

Bitcoin money is only a utility—not the thing of value in itself. It's a label. If bitcoins become too precious and scarce, there are always plenty of alternative blockchain currencies to use instead. Unlike the issuers of national fiat currencies, no one—not even the tax authority—is forcing anyone to use bitcoins. So they don't have the same role as the sort of money that was invented for early Renaissance monarchs to shut down the peer-to-peer marketplace. Amazingly, it's money people who have the hardest time understanding this part, which is why they are so destined to be burned on their bitcoin investments, however they play this one.

The only way to understand the real purpose and function of Bitcoin is to stop asking ourselves if it's a good investment. Even now, many of the people reading these words are trying to figure out whether I'm saying bitcoins will or won't be worth something: should they close the book and buy some right now, or not? If you need an answer in order to move on, then fine: Don't invest in bitcoins. You could make money if you buy them at the right moment, but that's not what they're for.

Although bitcoins need enough investor interest to prove their merit and gain acceptance with transactors, too much investor interest actually limits the currency's effectiveness. Bitcoin, as a program, is not meant to solve the problem of how people can invest money over time. It is addressing the problem of how people can transact securely without a central mediator and do so anonymously. And Bitcoin is most assuredly secure. For the record, the much-publicized bitcoin robberies and cyberattacks have been on some of the bitcoin exchanges and online wallet systems—one of them adapted from a gaming Web site that was never intended to secure banking records.[36] Even so, they have nothing to do with the Bitcoin blockchain itself, which is, for all intents and purposes, impenetrable.

Bitcoin's failure to overcome our business culture's bias for hoarding and scarcity may be a temporary setback, or it could prove to be a fundamental flaw in the way the system was designed. The Bitcoin blockchain generates an arbitrarily limited supply of bitcoins. It may have been meant to counteract what sometimes seems like the profligate pumping of money into the economy by central banks. But by setting a cap on how many bitcoins can ever be created, Bitcoin doesn't transcend the scarcity bias of central currency; it exacerbates it.

The only ones who don't think about bitcoin that way are the miners—those participants with the fastest computers—whose power to verify transactions and earn new coin puts them at the center of the economy, at least while new coin is being created. All the money originates with them, even though it leaves their hands when they spend it—with no strings or interest attached, unlike central currency. They may be a new kind of elite, but they're an elite all the same. They are the new bankers, even if they function from the periphery, and even if they exist only in the first ten years or so of Bitcoin's existence, until the money supply is completed.

Not that mining bitcoins is cheap. In addition to the hardware, miners must invest in a tremendous amount of computer processing and electricity consumption. In 2013, miners expended some 1,000 megawatt-hours per day verifying transactions and mining new bitcoins.[37, 38] As Bloomberg writer Mark Gimein noted, that's half the power needed to run the Large Hadron Collider. Less than one year later, PandoDaily placed the network's energy usage at 131,019.91 megawatt-hours per day, an increase of over 1,000 percent.[39] While specialized computers called "mining rigs" are improving the energy efficiency of bitcoin mining, the shrinking number of unmined bitcoins and increasing length of the blockchain are raising the level of computing power required to perform Bitcoin's proof-of-work problems.[40] So even if Bitcoin did turn out to be economically feasible, it is unlikely to prove environmentally sustainable.

Those with digital literacy, processing power, and the fewest qualms about wasting electricity have the advantage. Meanwhile, those who already have lots of money can simply rent bitcoin-processing computers from companies with names like LeaseRig to do mining on their behalf. Money still buys a seat at the table.

So Bitcoin takes capital creation away from bankers but gives it to programmers or those who pay them. It does fix some of currency's problem, in that it's no longer sourced for interest and no longer requires growth. But it's still scarce and hoarded and never stops taking from the physical environment. If anything, it's more like the original gold coin that proved too scarce to be practical than it is like the abundant and circulating market money that spawned the peer-to-peer economy of the bazaar. There's no central bank earning interest on the currency, but its value is still a product of its relative scarcity—the way cigarettes serve as cash in a POW camp. Money for prisoners. This creates a zero-sum mentality in its users and discourages circulation. There's only so much to go around, so it's better to hoard it than spend it.

Still, while bitcoins may ultimately prove to have limited value as a currency, the Bitcoin network represents a potentially epochal shift in how we organize finance, computing, corporations, and even our society. On even the most superficial level, thanks to the Bitcoin network, bitcoins are more verifiable than central currency and more collectively negotiated. The amount in circulation is entirely removed from the control of a Fed or a central bank, as well as from the politics and agendas that might inform their decisions. But on a more essential level, the Bitcoin protocol represents a profound leap in how we understand trust and security—two of the original functions of money.

Verification is no longer something we need from an outside authority. There is no official person or entity that can offer (or deny) a stamp of approval. Trust, safety, and ownership are guaranteed not by central command but by the network of participants. In this system, the power of a currency derives not from the enforcement capabilities of the central government but from the grassroots connectivity of the people in the marketplace. Money is not protected with bank vaults, real or virtual, but with the widest possible public oversight.

Wall Street speculators don't realize that (or perhaps don't want to realize it), because this change promises a radical shift away from the system by which they extract wealth through finance. Yet their willingness to bet on bitcoins, even in the wrong ways and for the wrong reasons, shows that the market does recognize there's a shift under way. They just don't know how to participate in it yet.

Bitcoin is actually bigger than money. The blockchain may not engender unilateral trust, but it compensates for our distrust of one another in a new way. Instead of an authority bearing witness to an agreement, we all do. There is no single "watcher" with a key to the ledger or a hand on the till, so the question of who is watching the watchers becomes moot. There is no authority—absolute or otherwise—to corrupt. Authority is distributed. If we look past all the cryptography, the algorithms, and the buying and selling, the blockchain is simply an open ledger—a collectively produced, publicly accessible record of agreements made between individuals. Additionally, it is verified by an anonymous peer group, then encrypted so that only those involved in the specific transaction know who participated. This has applications well beyond bitcoins.[41]

The blockchain can "notarize" and record anything we choose, not just the cash transactions between Bitcoin users. Entire companies can be organized on blockchains, which can authenticate everything from contracts to compensation. Decentralized autonomous corporations, or DACs, for example, are a fast-growing category of businesses that depend on a collectively computed blockchain to determine how shares are distributed. To count as a true DAC, a company must be an open-source endeavor whose operation occurs without the supervision of a single guiding body, such as a board or a CEO.* Instead, a project's governing rules and mission must emerge from consensus. Project workers are compensated for their labor or capital investment with shares in the blockchain, which increase in number as the project develops.[42] We can think of DACs as

* There's still some debate among participants over how to define DACs, Dapps (decentralized applications), and DAOs (decentralized autonomous organizations), as well as the principles to which they must adhere.

companies whose stock is issued little by little as the company grows from a mere business plan into a sustainable enterprise. Only individuals who create value for the company are awarded new stock proportionate to their contributions.[43]

Fittingly, the majority of DACs currently sell blockchain-related services themselves. By committing to the blockchain for their own governance and share distribution, DACs lend credibility to the technologies they are selling. They stand in stark contrast to the bitcoin ETFs being peddled by the Winklevoss twins and others, in which profit is extracted through traditional Wall Street markups and expense ratios, and transactions remain opaque. By using the blockchain, DACs subject themselves to total transparency. Everyone can see everything.

Even with all their advantages, there is a certain brittleness to most of these blockchain projects. Those who get in early tend to earn the most of whatever coin is being distributed. Moreover, the rules that get into a system in the beginning become pretty intractable. Unforeseen changes to the particular sector in which the blockchain is operating, or even to the whole world, are hard to account for on the fly. Finally, if everyone is supposed to keep a copy of the public ledger, that in itself can get pretty unwieldy once enough transactions have occurred; the Bitcoin blockchain alone is bigger than many people's hard drives. In short, as these real-time economy projects scale, their legacies become liabilities.

After all, no matter how promising the blockchain's applications may be, decentralized technologies don't guarantee equitable distribution; they merely allow for value to be exchanged and verified in ways that our current extractive, centralized systems do not. As we've seen, the Bitcoin project was intended only to address the issue of providing security through a decentralized ledger. It was solving for peer-to-peer verification with anonymity—not economic justice or even the healthier circulation of currency.[44]

Bitcoin does prove that there are distributed solutions to problems formerly considered the exclusive province of central authorities. We can do this ourselves. But its inherent anonymity does nothing to restore the human relationships decimated by corporate activity, and its functionality irrespective of distance does nothing to reassert the local realities in which human beings actually live.

To do that, we may have to turn not to a collectively negotiated digital file but to one another.

MONEY IS A VERB

As creatures of a digital age, our first impulse is often to apply some algorithm, computer program, or other technological solution to a problem. Bitcoin is just such an approach, turning the massive processing power of distributed personal computers to verifying the exchange of value. In using such a technology, we learn to trust the cryptocurrency's open-source algorithms over the bankers and authorities who may have abused that privilege in the past. In blockchain we trust.

Of course, the underlying assumption is that people can't trust one another enough to transact directly without the constant threat of double-dealing, fraud, or nondelivery of services. By implementing a money system that encourages us to put our faith in technology, we again usurp whatever social bonds our marketplaces may afford us. We tend toward an economic culture driven more by disconnection and brinksmanship than by bonding and mutual benefit. That's because we are influenced by the tools we use.

Behavioral economists know this all too well. Interestingly, as we saw earlier, it was only when debt-based currency's limits began to surface in the twentieth century that those legions of psychologists were hired by banks and credit companies to come up with ways to get people to borrow more money, and at higher interest rates. The psychologists learned how to exploit people's misconceptions about how money worked and gave names to each of our vulnerabilities, such as "irrationality bias," "money illusion bias," "loss aversion theory," and "time discounting."* Then, they

* People borrow not opportunistically but irrationally. As if looking at objects in the distance, they see future payments as smaller than ones in the present, even if they are actually

created products and wrote advertising that took advantage of these failings in order to get people to act against their own best interests. In other words, if the money stops fulfilling the needs of human beings, you change human beings to fit the needs of the money.

It's yet another example of the industrial-age ethos that places human needs and values below those of the greater machines and systems in which we live. By contrast, the digital media environment invites us to look at the systems we use as changeable programs, and people as the end users for whom those systems are built. Computer programs like Bitcoin may be the most explicitly digital expression of this drive to hack economics as if it were an operating system. But the Bitcoin protocols are still more concerned with replicating the functions of money than they are with serving the needs of humans. Indeed, the most far-reaching modifications of our debt and currency systems may turn out to be a lot less technological in their expression and more focused on the specific human problems they are attempting to address. They are not solving for money but solving for people.

For instance, one of the main issues that emerged in the wake of the Occupy Wall Street movement was the problem of debt. Student debt is estimated at about $1.2 trillion as of this writing, while medical debt currently burdens over fifty million adults in the United States[45] and is the nation's largest single cause of personal bankruptcy.[46] All these troubled debtors pose a problem for the debt industry as well: people in age groups that used to be counted on for buying first homes and taking out mortgages are still too busy paying back student loans to consider purchasing real estate, while homeowners facing medical debt are the leading cause of foreclosure and credit nonpayment.[47] Most banks and credit-card companies simply package and sell the bad debt to loan collectors and other bottom-feeders at pennies on the dollar, just to get it off their books. The debtors still owe the full amounts on their loans, but the new creditors

---

larger. They are more reluctant to lose a small amount of money than they are eager to gain a larger one—no matter the probability of either event in a particular transaction. They do not consider the possibility of any unexpected negative development arising between the day they purchase something and the day they will ultimately have to pay for it.

know they will collect on only a portion of it, if any, before the debtors go bankrupt or run away. Everyone ends up worse off.

During the Occupy Wall Street gatherings, activists considered the conundrum from every traditional approach. They looked at promoting resistance through collective debt refusal, forcing creditors to show the full chain of custody of loans on which they were trying to collect, or forming a PAC and lobbying Congress on banking and credit reform. But taking a cue instead from the hands-on, do-it-yourself bias of the digital age, the activists came up with a much simpler solution: buy the debt. They launched a project called the Rolling Jubilee,[48] raising money from donors to buy back and then dissolve debt. With just $700,000 of initial donations, they have managed to dissolve over $17 million of student debt and $15 million of medical debt and are now targeting payday loans and private probation debts.[49] And the more people they get out of debt, the more new donors they create.

Such solutions may not be highly technological, but they are digital in spirit, especially in the way they retrieve the peer-to-peer mechanisms of mutual aid and distribute personal risk and liability throughout a network. Finally, the solution itself is a hack of the existing, highly exploitative system; pennies-on-the-dollar leverage afforded to credit packagers is used instead to relieve debts twenty-five times greater than the donated amount. In the best cases, the benefactors can in turn donate that small amount required to bail them out, and the debt jubilee can keep rolling to absolve more debtors.

The remaking of the money system and its many debt-based tentacles can certainly make use of the net, social media, and blockchains. We have in consumer technology all the security and administrative capabilities that used to be the exclusive province of banks and major corporations. But the reprogramming of money requires less digital technology than digital thinking and purpose. As we put our fingers—our digits—to the purpose of a better money system, we must focus less on what we can accomplish with a particular set of technologies than what we *want* to accomplish with them. With those goals clear in our minds, we can evaluate the solutions on offer or develop new ones of our own.

So instead of asking how we can manipulate human financial behavior to serve existing forms of money, we ask: What sorts of money will encourage the human behaviors we admire and long to practice? What sorts of money systems will encourage trust, reenergize local commerce, favor peer-to-peer value exchange, and transcend the growth requirement? In short, how can money be less an extractor of value and more a utility for its exchange? Less prone to getting stuck in capital, and more likely to remain dynamic and flowing?

1. Local Currency

The simplest approach to limiting the delocalizing, extractive power of central currency is for communities to adopt their own local moneys, pegged or tied in some way to central currency. One of the first and most successful contemporary efforts is the Massachusetts BerkShare, which was developed to help keep money from flowing out of the Berkshire region.

One hundred BerkShares cost ninety-five dollars and are available at local banks throughout the region. Participating local merchants then accept them as if they were dollars—offering their customers what amounts to a 5 percent discount for using the local money.[50] Although it amounts to selling goods at a perpetual discount, merchants can in turn spend their local currency at other local businesses and receive the same discounted rate. Nonlocals and tourists purchase goods with dollars at full price, and those who bother to purchase items with BerkShares presumably leave town with a bit of unspent local money in their pockets.

The 5 percent local discount may seem like a huge disadvantage to take on—but only if businesses think of themselves as competing individuals. In the long term, the discount is more than compensated for by the fact that BerkShares can circulate only locally. They remain in the region and come back to the same stores again and again. Even if nonlocal stores, such as Walmart, agree to accept the local currency, they can't deliver it up to their shareholders or trap it in static savings. The best Walmart can do is use it to pay their local workers or purchase supplies and services from local merchants—again, supporting the local economy instead of absorbing those externalities.

Simple, dollar-pegged local currencies like BerkShares are depending on what is known as the local multiplier effect.[51] Money of any kind, even regular old dollars, spent at local businesses tends to stay within the local economy. That's because local, independent businesses tend to source their materials and services from nearby instead of from some distant corporate headquarters. According to a broad study conducted by the American Booksellers Association, 48 percent of each dollar spent at locally owned retailers recirculates through the community, compared with 14 percent at chain stores.[52] With geographically limited local currencies, that number stays close to 100 percent, until they are exchanged back into dollars. Such currencies are biased against extraction and toward velocity.

BerkShares can purchase all the quaint New England commodities one would expect: local produce, a cup of fair trade coffee, a stay at the bed and breakfast. But they are also exchangeable for more utilitarian goods and services: construction contracting, Web design, even a trip to the undertaker.[53] That's where it gets tricky. Local currencies work best for locally generated goods and services, or when a commodity's markup is derived from a locally added value, such as atmosphere or labor. They don't work as well for selling goods that are at the end of long supply chains or depend on commodities sourced from far away. A contractor can use BerkShares to buy locally produced shutters, but he can't buy his tools or nails with them. A bookshop owner can offer her customers an effective 5 percent discount on their purchases, but she's still paying full price for her inventory. It's just not in a local business's short-term interests to accept local currencies. And the multiplier effect really works only if there are a whole lot of businesses willing to play along.

Local-currency advocates acknowledge this shortcoming but believe it is mitigated by a currency's visibility. With geographically based currencies, the thinking goes, the "buy local" ethos becomes visible—still voluntary but validated by merchants and political leaders. Unless you're spending BerkShares, how can you make it clear to merchants that you are thoughtfully supporting local business? Local currencies are their own best publicity, rendering "buy local" visible and thereby fostering the community spirit and soft peer pressure that lead to widespread buy-in and network effect. Then again, some customers might demonstrate their loyalty to local merchants by forgoing the local currency discount altogether and spending "real" money without the discount.

Many other communities are experimenting with variations on the BerkShare model. Proponents claim that by being removed from the greater economy, these currencies work against the scarcity bias of central currency and are more resistant to boom, bust, and bubble cycles.[54] Detroit Dollars, Santa Barbara Missions, and, in the UK, the Bristol, Brixton, and Cumbrian Pounds each offer their particular variations. Detroit Dollars offer much the same arrangement as BerkShares, only at a 10 percent discounted exchange rate.[55] The UK's Bristol Pound is backed by a credit union, has a digital debit payment system, and can be used by businesses to pay certain taxes. A pilot program in Nantes, France, promises to allow citizens to pay municipal fees in local currency.[56]

Most of these local currencies are still more fad than utility. In some cases, it's because only economically conscious progressives are willing to employ them, and then only for goods and services that can't command clear value on the regular open market, such as spiritual healing or career counseling. In others, it's because the central monetary system is still strong enough to serve a majority of people—or dominant enough to minimize the apparent viability of alternatives. Moreover, by pegging themselves to central currency, these local discount currencies can isolate themselves only so much from the chronic monetary problems of inflation, deflation, bubbles, and debt.

2. Free Money: Cash as Utility

For a local currency to distinguish itself as more than fashion, it must be utterly independent of existing money systems—that is, pegged to nothing but itself. Maybe it's only when an economy gets truly bad and money is nowhere to be found that more bootstrapped applications of alternative currency tend to surface.

For example, after Germany's defeat in World War I, much of the German-speaking world was in economic shambles. In the Austrian city of Wörgl, over 30 percent of workers were unemployed and a significant portion of the population was destitute.[57] There was not enough currency—Austrian schillings—to go around, not with Austria paying off its war debt to the banks.

Inspired by the "free money" theory of German economist Silvio Gesell, the mayor of Wörgl decided to create a local currency programmed to solve this particular crisis. According to Gesell, the core qualities of a money system—its biases, the way it extracts value or discourages circulation—are unknown to the people using it until they are shown an alternative. Gesell was no Marxist; he was a free-market advocate but particularly antipathetic toward charging interest for money, which he believed was the way that moneyed classes prevented others from participating fully in the economy.

The town had plenty of workers and plenty of resources, plenty of needs and plenty of providers; it just didn't have a means for all those people and businesses to exchange goods and services. Whatever money they were capable of generating went back to servicing debt. So the mayor took the town's entire treasury of 40,000 Austrian schillings and put it in the local savings bank as a partial reserve against a new kind of currency: labor certificates that came to be known as Wörgl.[58]

He got the town working again by paying people in Wörgl to attend to public works projects, such as roads and schools. Because he had read Gesell, he made sure to tilt the currency toward local prosperity. By design, the labor certificates functioned as "circulation only" currency. Citizens could use them for goods and services and even to pay their local taxes. But Wörgls were terrible for saving: the certificates lost value, "demurring" at a rate of 1 percent per month, much like a grain-based currency of the late Middle Ages. As an unexpected consequence, this demurrage incentivized citizens to pay their taxes early, which in turn gave local government the liquid assets it needed to manage infrastructure and create still more employment projects. Like the Rolling Jubilee, the new currency initiated a virtuous cycle. In the midst of global depression, the village built bridges, homes, a reservoir, even a ski jump.[59]

By programming the Wörgl as an antigrowth currency that lost value over time, the mayor discouraged hoarding, freeing money to function in its needed role as a medium of exchange. The design privileged investment in local development ahead of servicing runaway debt. In the thirteen months of its existence, the Wörgl's tremendous success drew a little too much attention from the wrong quarters. Viewing it as a threat to its monopoly, the Austrian Central Bank declared the Wörgl illegal. All Wörgls were removed from circulation, scarcity returned, and unemployment rose back to its peak rate.

There were many other successful local currency trials in Austria and Germany, and they were all met with a similar response from central lawmakers. Since they could not be used outside their respective regions, the local currencies had no value to the centers of political and economic power. They were stabilizing to real people in real places but destabilizing to those who sought to centralize control over Germany. If anything, the prolonged and unnecessary depression merely paved the way for the discontent that fueled Fascism.

Free local currencies were also responsible for providing a means of transaction during the Great Depression in the United States. Some were successful enough to pose a threat to central powers; others were merely successful enough to get traditional banking running again. In a much more pragmatic set of writings than those of Gesell, Yale University economist Irving Fisher argued that the sole focus of an alternative currency in such circumstances should be to increase the velocity of money.[60] He advocated the use of "stamp scrip" as a weapon against deflation.[61] Stamp scrip would come with the requirement that it be spent and stamped at regular intervals in order to maintain its value. Only after it had been fully stamped—meaning it had been spent thirty-six or fifty-two times, depending on the particular note—could it be redeemed at the bank. The money therefore had a built-in incentive to be spent. And it worked, at least within the local communities that used it.

Another depression-era currency, "tax anticipation scrip," was issued by a few dozen American cities from Ann Arbor, Michigan, to Tulsa, Oklahoma,[62] whose municipal funds had been lost in the banking crisis. Without any money, these city governments began to pay their workers and suppliers in small-denomination IOUs against future tax revenue. The scrip usually circulated at a discount of its face value, but that was better than nothing for workers and citizens of these otherwise bankrupt economies.

In perhaps the most straightforward solution to scarce currency, many depression workers joined barter exchanges and self-help cooperatives. Sometimes, workers would exchange their labor with factory owners for some of the goods they produced, or with farmers for a share of the crops. In other systems, laborers swapped goods and services directly with one another. They used various forms of scrip simply to keep track of how much work people had put in and how much value they had taken out. The founders and users of these currencies were not Communists or even ideological. Many, including Organized Unemployed Inc.'s Reverend George Mecklenburg, abhorred state aid so much that his sauerkraut co-op's scrip was imprinted with the slogan "Work, Not Dole!"[63]

As FDR's federal government programs kicked in, these alternative local money systems were either declared illegal, abandoned, or used to prime the larger economy. In many cases, though, they gave communities a taste of self-sufficiency before aid in the form of new debt arrived. There's no doubt that FDR's programs, bonds, mortgage policies, and GI bills let more Americans live better. He successfully forced the banks to start lending, which set us on the path to seven or eight decades of mandatory currency expansion and economic growth—for better and for worse. If lending can be understood as a form of dole, then Mecklenburg's slogan may be more prescient than he meant it.

3. Cooperative Currencies: Working Money into Existence

Today, economic recovery is still understood as big institutions lending other institutions a bunch of money in order to capitalize new development. When cities have no cash and unemployment is high, the governor or president is supposed to incentivize a bank to lend money to a corporation to build a factory to create some jobs to kick-start the economy. Eventually, though, everyone up the chain of capital has to be paid back, the scores of financial advisors and other stakeholders gaming the loans get to take their cut, and the municipality is left dependent on a foreign corporation for everything. The corporation has the leverage to demand better tax treatment as it extracts more value from the region than it creates. Then the company leaves, and the cycle begins again. And that's considered a success story—the very premise of our cyclical economy.

It would be much simpler, more sustainable, and less expensive to get that region to work without putting it into debt or the service of a remote entity. Instead of installing industry, government could much more easily equip regions with the tools and information they need to develop a means of value exchange.

After all, if people have skills and needs, then they have the basis for an economy. All they require is a way of exchanging value with one another. If Joe fixes refrigerators, Mary bakes bread, Pete grows wheat, and Sylvia babysits, each one can support the others. But barter alone won't do it. Sylvia may need her refrigerator fixed, but Joe has no kids for her to babysit. He needs bread. But Sylvia can't babysit for Mary, whose kids are off at college. Sylvia can babysit for Pete. But how does that help?

If everyone had dollars, it would be easy. Pete would pay Sylvia to sit for his kids. Sylvia would pay Joe to fix her fridge. Joe would pay Mary for bread. And Mary would pay Pete for the grain to make her bread. If only Pete had some cash, he'd be able to hire Sylvia. So he votes for the politician who promises new capital investment, a new plant, new jobs, some new dollars, and ultimately a new set of loans to be paid back to the bank for supplying the money to begin with.

What if, instead, Pete and the rest of his community learned to transact directly? What if, instead of bribing foreign business with a tax-free new enterprise zone, the politician simply offered a PDF file with simple instructions on how to start up a "favor bank" or local currency? Or maybe offered up an advisor for a couple of weeks to the local chamber of commerce to get an exchange system or a new community currency off the ground? Unlike local discount currencies, such as BerkShares, cooperative community currencies don't need to be pegged to the dollar at all. They are not purchased into existence but are worked into circulation—a bit like the market money of the late Middle Ages. They are best thought of less like money than like exchanges.

The simplest form of cooperative currency is a favor bank, such as those being employed in Greece and other parts of southern Europe during the euro crisis. Incapable of finding work or sourcing euros, the people in many places lost the ability to transact. Even though a majority of what they needed could be produced locally, they had no cash with which to trade. So they built simple, secure trading Web sites—like mini-eBays—where people offered their goods and services to others in return for the goods and services they needed.[64] The Web sites did not record value amounts so much as keep general track of who was providing what to the community and coordinate fair exchanges. This casual, transparent solution works particularly well in a community where people already know one another and freeloaders can be pressured to contribute.

Larger communities have been utilizing time dollars, a currency system that keeps track of how many hours people contribute to one another. Again, a simple exchange is set up on a Web site, where people list what they need and what they can contribute. The bigger and more anonymous a community, the more security and verification is required. Luckily, dozens of startups and nonprofit organizations have been developing smartphone apps and Web site kits through which local or even nonlocal communities can establish and run their own currencies.[65] Time dollars can even be run on a blockchain, where the provider and the purchaser verify the transaction.

Time exchanges tend to work best when everybody values their time the same way or is providing the same service. The Japanese recession gave rise to one of the most successful time exchanges yet, called Fureai Kippu, or "Caring Relationship Tickets." People no longer had enough cash to pay for their parents' or grandparents' health-care services—but because they had moved far away from home to find jobs, they couldn't take care of their relatives themselves either. The Fureai Kippu exchange gave people the ability to bank hours of eldercare by taking care of old people in their communities, which they could then spend to get care for their own relatives far away. So a girl might provide an hour of bathing services for an elder in her neighborhood in return for someone preparing meals for her grandfather who lives in another city. As the Caring Relationship Tickets became accepted things of value, people began using them for a variety of services.[66]

What's more, as the Japanese economy recovered enough for people to afford health care from traditional for-profit providers, a majority elected to stay within the Fureai Kippu system. Not only was it less expensive for young people to pay directly with their hours, but the elders found their amateur caregivers more connected and compassionate.[67]

Different money systems engender different behaviors and attitudes from the communities that use them. With Fureai Kippu, the caregivers get a surrogate elder for whom they can provide the care they wish they could be providing their own grandparents. They can establish relationships and negotiate scheduling directly, without having to worry about their company's profits, schedules, billing, or insurance—or the costs of all those things. People can work less and get more. It's only a problem for those who would hope to extract value from the transaction.

Time-dollars systems, and those like them, don't encourage extractive profiteering because they are really only good for exchanging labor and services with other people. Unlike bank-issued currency, hours are not borrowed into existence, nor do they collect interest over time. They don't actually accumulate. Instead, everyone's account starts at zero. When Sylvia babysits for Joe's kid, her account is credited three hours, and Joe's is debited three hours. Since they both started at zero, Sylvia now has three hours in her account, while Joe's account is in the red for three hours' worth of work. He will remain owing those three hours to the system—to the community—until Mary hires him to fix her refrigerator. The net result of the exchange is even. The net total of the system is still zero. It is not a growth economy but a transactional economy.

Although a person can do a bunch of work in order to bank enough hours to get a whole bunch of services, most time exchanges put a limit on how many hours members can accumulate. They also put a limit on how many hours a person can owe. This way a freeloader can be removed from the system, and the entire community can absorb the cost of the unearned hours pretty easily.

The time banks of progressive communities such as Ithaca, New York, or Boulder, Colorado, are some of the most famous in existence, but they have also begun to spring up in depressed regions all over the world, thanks largely to the ease with which they can be administered online. For instance, throughout Spain, time banks of all shapes and sizes are offering people a way to work and trade independent of a monetary emergency that the average Spaniard had no part in creating. There are more than a hundred time banks in Barcelona alone, with membership sizes varying from less than fifty into the thousands. Some of them are managed day-to-day by human staff (who are themselves compensated in time dollars), while others are entirely digital and automatic. Many of the bigger time banks even offer checking, auditing, and online banking.[68]

Time dollars are extremely egalitarian, valuing each person's time the same as anyone else's. An "hour" is worth one hour of work, whether it is performed by a plumber or a psychotherapist. Another version of time dollars, called LETS (Local Exchange Trading System), allows people to negotiate the value of their own hours or services. The advantage to a LETS is that it can also account for the cost of goods or even the capital investment of training required for certain services. In a purely time-based system, an auto mechanic can charge "hours" for the time he spends fixing a car, but he must charge standard currency for the parts. A LETS allows him to set a local value for the parts he has bought.

Of course, eventually the mechanic will need some regular dollars if he is going to stay in business. Auto parts manufacturers don't accept local scrip. That's why time and LETS systems are best thought of as complementary currency systems rather than complete replacements for all forms of central money. They *complement* central currency, giving people a great way to conduct local business when cash is either too scarce or too expensive. Moreover, they encourage transactional velocity and even good will instead of hoarding and profiteering. More people get to do more things for one another.

A local currency needn't be the right tool for every economic purpose in order to be successful. A LETS won't be getting people new iPhones, snowblowers, or iron ore anytime soon. Complementary currency's purpose, more often than not, is either to kick-start a local economy or to make local transactions less burdened by the cost of currency and thus more competitive with nonlocal corporate, chain store, or big-box offerings. If a local farm and a local biodiesel company become members of a LETS, then the community has a great majority of what it needs to survive, even if it's got no money at all.

The more flexible structure of a LETS makes it great for some of the more hybrid approaches to economic renewal and sustainability. By tying LETS dollars to something else in the real world, users can more easily gain trust and traction. For example, Mayor Jaime Lerner used a modified LETS to address both the economic and environmental crises facing Curitiba, Brazil. Sanitation and pollution in the slums were out of control, but Lerner was leery of requesting large loans from sources such as the World Bank, which he knew could well result in a permanent state of debt.[69] The city slowly undertook a modest program aimed at ameliorating its sanitation problems, which grew into one of the most successful, long-running LETS on record. It began by offering children bus tokens in exchange for collecting garbage. The children, many of whom were too poor even to go to school, helped clean up the shantytowns, thus supplementing the work of an overburdened sanitation department. As the children brought their tokens home, their parents were able to use them to go into the city to look for work. Soon, merchants began to accept the tokens as normal currency, administered as a LETS.

Once this happened, Lerner had the freedom to expand the program even further, using the bus-token-based LETS currency to pay for public works projects, such as housing and infrastructure. The LETS was allowing him to invest in the city and its future instead of simply ameliorating poverty.[70] True, this ended up putting more bus tokens into circulation than were actually needed for transportation purposes, but by then the value of the tokens on the LETS exchange meant more than the seats on the bus that originally backed them. Twenty-five years later, the city has some of the highest quality-of-life indices in the world. Approximately 70 percent of households still participate in the garbage-for-tokens program, and the city is able to recycle some 60 to 70 percent of its trash locally. Today, the United Nations points to Curitiba as a leader in sustainable urban development.[71] The Curitiba bus token shows that LETS are scalable and needn't be limited to small communities. Once a region agrees upon a medium of exchange, it can enjoy the liquidity of monetary stimulus but with no day of reckoning when the central bank comes to collect. The region is not expected to grow just to pay back interest; it can simply recover and reach sustainability.

Complementary currency helps to insulate communities disempowered by deflation, inflation, or volatility as the economy moves from one extreme to another. All too often, the financial crises that merely unsettle the hubs of political and economic power truly decimate outlying regions and smaller communities at the very end of the money supply chain. Local currencies, time dollars, and LETS are programmed to be nonextractive, highly transactional, and free of borrowing costs.

4. From Extracting to Enabling: The New Local Bank

In such a climate, how are traditional banks to participate effectively in the financial rehabilitation of the communities they serve? How are they supposed to function at all once people begin to see central currency as extractive, and the bank as a foreign corporation removing value from the community in the form of interest? Again, just as for their corporate brethren, it's by adopting a hybrid approach. Here's just one possibility:

Sam's Pizzeria is thriving as a local business, and Sam needs $200,000 to expand the dining room and build a second restroom. Normally, the bank would evaluate his business and credit and then either reject his loan request or give him the money at around 8 percent interest. The risk is that he won't get enough new business to fill the new space, won't be able to pay back the loan, and will go out of business. Indeed, part of the cost of the loan is that speculative risk.

In a more hybrid approach, the banker could make Sam a different offer. The bank could agree to put up $100,000 toward the expansion project at 8 percent *if* Sam is able to raise the other $100,000 from his community in the form of market money. Sam is to sell digital coupons for $120 worth of pizza at the expanded restaurant at a cost of $100 per coupon. The bank can supply the software and administrate the escrow. If Sam can't raise the money, then it proves the community wasn't ready, and the bank can return everyone's money.

If he does raise the money, then the bank has gained the security of a terrific community buy-in. Sam got his money cheaper than if he borrowed the whole sum from the bank, because he can pay back the interest in retail-priced pizza. The community lenders have earned a fast 20 percent on their money—far more than they could earn in a bank or mutual fund. And it's an investment that pays all sorts of other dividends: a more thriving downtown, more customers for other local businesses, better real-estate values, a higher tax base, better public schools, and so on. These are benefits one can't see when buying stocks or abstract derivatives. Meanwhile, all the local "investors" now have a stake in the restaurant's staying open at least long enough for them to cash in all their coupons. That's good motivation to publicize it, take friends out to eat there, and contribute to its success.

For its part, the bank has diversified its range of services, bet on the possibility that community currencies will gain traction, and demonstrated a willingness to do something other than extract value from a community. The bank becomes a community partner, helping a local region invest in itself. The approach also provides the bank with a great hedge against continued deflation, hyperinflation, or growing consumer dissatisfaction with Wall Street and centrally issued money. If capital lending continues to contract as a business sector, the bank has already positioned itself to function as more of a service company—providing the authentication and financial expertise small businesses still need to thrive.

The bank transforms itself from an agent of debt to a catalyst for distribution and circulation. Like money in a digital age, it becomes less a thing of value in itself than a way of fostering the value creation and exchange of others.

Less a noun than a verb.

Chapter Four

INVESTING WITHOUT EXITING

"Our favorite holding period is forever."

—Warren Buffett

FINANCE IS NOTHING PERSONAL

Okay, suppose the digital economy delivers on its deepest promises. Let's say we transcend our industrial-age thinking and develop platforms dedicated to serving the needs of people instead of simply removing them from the value equation. Let's say we transition to a business landscape in which corporations don't have to grow in order to remain solvent. And, as long as we're doing so well, let's say we get down to the operating system of money itself and reprogram it from the ground up to be biased less toward preserving passive wealth for the rich and more toward exchanging value among everyone else.

Retrieving the ethos and mechanics of the medieval bazaar may be great for promoting live trade, but how does it provide stability or security? When everything is moving, how does anyone accumulate wealth? If money really is about to become less of a noun than a verb, then what's an investor supposed to do?

Sure, it's easy for the 99 percent to celebrate the impending collapse of the investment economy; we've bailed out the private bankers with public funds enough times that we can't help but delight in the fact that they're losing their ability to extract value out of thin air while the rest of us work for a living.

And believe me, they are scared. The finance and investing conferences I've spoken at over the past decade have increasingly been characterized by panic: Interest rates of all kinds are too low to make much money collecting them. Holding money doesn't work in a low-interest environment, so investors are forced into stocks, commodities, and private equity. And with everybody crowding into those investments, they end up overvalued in comparison to the underlying assets.

Central bankers pour more money into the economy, and they scratch their heads in amazement that it doesn't lead to jobs and inflation. What they don't see is that employment and prices are remaining low because the injected money isn't trickling down into the economy of wages and products. It stays in the investment sector, where it's the financial instruments themselves that are inflating. Meanwhile, wealth managers and hedge fund operators I encounter are less spooked by inflation than they are by *de*flation. As we get more efficient at sourcing and making stuff, prices go down, profits go down, and wages go down—but all the investment capital still needs to park itself somewhere and, ideally, grow.

It's not just the bankers' problem. Although most of us aren't running hedge funds, we, too, need a place to save effectively for our kids' college educations and our own retirements. Favor banks and LETS systems are great for those who are willing to work and keep working, but does escaping the growth trap mean that regular people lose any ability to invest and accumulate capital for the long term?

Initially, it appeared that digital technologies would be a boon for the individual investor—the little guy. The net would make markets more transparent, spread financial information more democratically, and let people trade for themselves, just like the pros. It should provide a way to disintermediate the bankers and brokers—cutting out the middlemen and keeping more autonomy and cash for oneself. Digitally powered trading seems to be further along an ever-improving continuum of agency for individual investors. But for all this to be true, we have to accept the underlying premise that individuals taking charge of their own future security is a winning proposition.

Unfortunately, the best evidence indicates that it's not, nor was it ever really meant to be. In *Forbes* writer Helaine Olen's analysis of the finance industry, *Pound Foolish*, she reveals that our current retirement savings practices have less to do with making secure futures for ourselves than with growing a profitable new sector of financial services around this human need.[1]

For most of the postwar period, Olen explains, American workers enjoyed the promise of employer-provided pensions. The companies they worked for would garnish a percentage of every employee's paycheck, invest it, and return an agreed-upon sum, beginning when the worker retired and continuing until he died. If its return on investment was greater than its pension commitments, the company could keep the surplus as profit. By the same token, if its pension fund failed to generate high enough returns to meet its benefit commitments, the company was obliged to make up the difference. That missing money had to come from somewhere, and neither executives nor shareholders appreciated spending profits on past employees, who had long since ceased providing their value.

Of course, in most cases companies found creative ways of passing the liability on to their employees, anyway. They would renegotiate pension commitments[2] or pit union members against one another by offering to pay for retirement benefits of the old by reducing those of junior employees.[3] A dismal pension outlook could even give a company the leverage it needed to reduce salaries and benefits across the board,[4] threaten bankruptcy, or move to a right-to-work state with no union at all.[5]

Many companies even abused the pension system, dipping into

supposedly protected pension accounts to fund ongoing operations or risky investments. The worse things got for a company, the more likely it was to see the pension fund as an emergency bank account. In some instances, companies would default altogether on their pension obligations after investing in risky or self-serving instruments.*[6]

In the economic downturn of the 1970s, many pension funds failed to meet their required targets, leaving companies on the hook for the difference. So they began to look for a way to relieve themselves of any fiduciary responsibility for their employees' retirement. The whole idea of retaining employees as quasi-family members over the course of their entire careers and then rewarding them with money for life seemed quaint but contrary to the free-market principles of the Thatcher-Reagan era, anyway. Companies and government alike began to treat employees more as independent contractors—free-market players, personally responsible and ultimately dispensable.

Independent contractors and other self-employed workers were already using tax-deferred individual retirement accounts to save for their futures. Although IRAs rarely made as good returns as professionally managed and collectively pooled pension funds, they were still a legal instrument through which self-employed people could earn tax credits for making contributions to their own retirements. Then, in 1981, a benefits consultant named Ted Benna looked at the existing tax codes and argued that they could be read in a way that would allow companies to provide workers with the opportunity to invest in IRAs.[7] Businesses jumped on this interpretation and ran with it to Congress.

Luckily for them, their new legislation suggestions promoting individual financial responsibility dovetailed perfectly with the greater Reaganite theme of personal empowerment. People, it was believed, should have direct access to personal finance, the game of speculation. And so,

---

* That's what happened in the famous Enron scandal, when the energy company used pension funds as part of its illegal investment scheme, costing employees $2 billion of their own retirement savings.

with the help of a salivating financial services industry, the now-ubiquitous 401(k) was legislated into existence.[8]

A hybrid of the IRA and traditional pension plans, the 401(k) is funded by a monthly percentage of the employee's salary and, sometimes, an optional contribution from the employer. The employee then picks from a range of diversified portfolios, administered by an outside financial firm. Instead of guaranteeing a lifetime of benefits, employers now only have to provide access to a plan and a matching contribution up front, if that. The employee alone is responsible for whether the plan appreciates, keeps up with inflation, or is invested responsibly. Moreover, the costs of asset management, brokerage fees, and financial services are also shifted from the company to the individual employees.[9]

The finance industry loved this new product, since it meant that instead of advising a few pension-holding companies, it could advise millions of new 401(k)-holding individuals. By saving individually instead of collectively, workers end up paying hundreds or even thousands of times for the same advice. Plans can charge a base fee upwards of 0.65 percent—even before kickbacks from brokerage houses and internal fund fees—making it hard for any of these plans to accrue earnings, much less keep up with inflation.[10] Those charges apply whether the fund makes or loses money. Neither the financial advisor nor the plan administrator is liable for results. Under the new scheme, a much larger portion of the same pot of retirement money could be extracted in fees to support the careers of many more financial advisors and services, since now everyone gets a customized, personal account.

Financial firms also won a vast pool of new clients with very little financial acumen and no real bargaining power—a far cry from the professional, corporate pension managers of the past. The industry made every effort to market retirement plans as tools of empowerment for individuals. As their own marketing research shows, however, they were actually pitting the unique weaknesses of individual investors against themselves, leveraging the investors' ignorance of the marketplace and its rules, as well as known gaps—what gamers would call "exploits"—in people's financial psychology.[11] The more that financial firms promoted these plans, the more employers were free to drop their pensions and the more workers came to rely exclusively on their own savings plans and market skills. This channeled additional money into the finance industry, which then had funds to spend on marketing for more profitable financial products and on lobbying for less regulation in creating them.

In our digital society, we take for granted that retirement is one's personal responsibility. Young professionals understand that they're playing a game, competing against one another in the marketplace of jobs as well as that of retirement strategies. As the United States' manufacturing base declines, fewer young workers expect old-fashioned, long-term guarantees such as pensions, anyway.[12] The rise of the 401(k) and concurrent decline in pensions emerged at a propitious moment in American history, when a strain of "free market" fundamentalism had seeped from the Goldwater and Friedman fringes of the Republican Party into the techno-libertarian mainstream. The long boom of the 1990s, and its accompanying corporate focus on lean management and cost cutting, only amplified that trend.[13]

This confluence of circumstances—the invention of individual, tax-deferred savings instruments, the decline in American manufacturing, the rise of ultrafree-market ideology, the growing power of finance as a PR and lobbying force, and the availability of online trading tools—created a feedback loop in which each element further exaggerates and entrenches the others. Between 1979 and 2012, pension enrollment rates dropped from 28 percent to 3 percent of employees.[14]

All this would be fine if individual retirement accounts performed as well as or better than the old pension plans. But as reported by sources from *Forbes* to *USA Today*, 401(k) participants actually end up saving *less* money, not more—and certainly not enough to retire on securely. Managing and monitoring retirement saving accounts require a degree of financial acumen that is simply beyond that of the average person. (It's actually beyond the capability of most advisors.) Commissions and financial fees—often obfuscated—account for the rest of the decline in returns. In fact, until the hard-won (but easily and regularly rescinded) banking reforms of 2012, retirement fund managers were not even required to report the fees they charged.[15] That's right: from the emergence of these plans in 1981 until the summer of 2012, there literally was no legal requirement to inform consumers how much they were paying for the privilege of having a retirement account. They could look up the fees internal to the mutual funds, but the financial firms administering the accounts were not required to disclose them. And consumers were accordingly clueless. A 2011 AARP survey noted that 71 percent of 401(k) holders erroneously believed that they were not being charged any fees, while another 6 percent admitted they did not know whether they were being charged or not.[16] As a result, according to *Forbes*, in 2011 pension-style plans performed at a 2.74 percent rate of return, while 401(k)s actually *lost* an average 0.22 percent.[17]

So personal retirement plans are sold as a means of empowering the individual investor to get in on the game, but in practice they more frequently exploit a person's ignorance and lack of negotiating power.[18] For a middle-class worker's 401(k) to perform well enough to retire on, he must not only invest like a pro but also never get seriously ill, never get divorced, and never get laid off.[19] In other words, it doesn't work in real life.

By tasking individuals with their own retirement savings, companies transferred risk to employees, shifted profits to shareholders, and created a tremendous new market for financial services—which in turn siphoned off more value from people to the banking sector. Nevertheless, a majority of us still hold out hope that those upward graphs and pie charts about compound earnings are really true—even though our investments are not going up as advertised. Most of us believe the story told to us by our employer-assigned financial advisors and the business press: that over time, those of us keeping our money in the stock market will average 7 or 8 percent a year. MIT economist Zvi Bodie has looked at the composite of the S&P since 1915 and shown the true rate of return for any forty-four-year period of investment over the past century. The real average is about 3.8 percent. The best moment for a person to have retired would have been 1965, for an average gain of about 6 percent. Retire today, and you're looking at having made under 3 percent.[20] Before fees, of course. So much for taking charge of our own finances.[21]

Fully aware of these liabilities, the financial services industry became less concerned about helping people invest for their own futures than about finding ways to make money off that very need: to game the system itself, all the while finding new ways to make investors feel they were getting in on it. This makes the typical pyramid scheme of the original stock market look honest by comparison. On the stock exchange, at least, those who get in early can win—albeit at the expense of those who come in later. That's just the way investing works. You speculate on the future value of things—buy low, sell high. The patsy is the guy doing the opposite. In the 401(k) game, the patsy is anyone who follows the advice of the human resources department and surrenders a portion of his or her paycheck to the retirement planning industry, all under the pretense of personal responsibility.

Extending this trend, digital trading platforms require individual investors to take even more responsibility for their own investments and correspondingly shoulder yet more risk. So digital trading is not an improvement; it's just one step further along the continuum from zero-maintenance, guaranteed pension pools toward highly participatory, individuated, and competitive investing.

To be sure, the emergence of online trading utilities gives retail users unprecedented access to previously obscure and inaccessible markets. Before online trading, human stockbrokers used to stand between people and any trades they wished to make. A client would call his broker on the phone, ask for information, make a decision based on what he was told, and then instruct the broker to execute the trade. That broker would, in turn, phone a trader on the actual floor of the exchange, who would walk up to a human specialist, who would match a seller with the buyer and then execute the transaction. Of course, every human in the trading chain either skimmed a bit off the top or placed a bit of margin between the buy and sell prices to collect as commission. They extracted value, but their very existence in the process also slowed things down to a human pace and worked to keep things from getting too far out of hand. Anxious shareholders could be talked down off the ledge by their brokers, and panic selling could effectively be halted or counteracted by a skilled specialist.

In the 1980s, discount brokerages emerged that allowed customers to call toll-free numbers and place trades with operators instead of brokers. Eventually, customers could even use the buttons on their phones to hear twenty-minute-delayed stock quotes and place orders based on this seemingly fresh information. This was the first form of interactive trading, and it led to a huge increase in volume and commissions for the brokerage houses executing the trades—with some, Charles Schwab for instance, reporting volume spikes of 35 percent or more.[22] As volume increased, brokerages created their own trading desks. Exchanges became more decentralized, automated, and virtual, making the human specialists who used to maintain an orderly market less and less relevant.

The proliferation of the Web took all this even further, as discount brokerages competed to provide their clients greater levels of access and control through online trading platforms. As early as 1999, 25 to 30 percent of all trades were taking place on sites like E*Trade, TD Waterhouse, and Schwab.[23] While online services did give retail traders more access to published stories and eventually even real-time stock quotes, they didn't genuinely increase traders' participation in the deeper game of the stock market. The information was still old and superficial compared with what the professionals knew, and the trades themselves were still dependent on brokerage houses with official seats on the exchange or, in other cases, enough of their own traffic to trade clients' shares internally.

Moreover, the platforms and utilities offered to retail online consumers resemble those of professional traders but don't really work the same way or offer the same levels of access. Programs such as E*Trade's "Power E*Trade," for example, break up the screen into several smaller panes with changing numbers, tickers, and flashing red and green boxes. They look like miniature versions of the computer setups of professional traders. But, unlike the tools used by the pros, they do not offer what is known—in lingo that sounds as if it were pulled from a video game—as Level 2 or

Level 3 data. These higher levels of data are derived from the relationship of the traded price to the spread between the buy and sell, and help traders determine the actual buying or selling pressure on a particular stock—which is a great advantage over those who see only the price or its movement.

Even traders who understand that such tools exist and pay the fees to access them are not trading directly with others. This is not the control panel for a trading desk. There's no one on the other side of the screen. It's just a professional-looking interface on an entirely retail amateur platform, connecting individuals to the computers of the same old brokerage house. There is still no peer-to-peer activity going on, through which empowered Internet users can bypass profit-taking intermediaries.

The gamelike control panels simply offer a more participatory *experience* of trading—along with positive feedback for making trades—which in turn encourages more activity. By 2000, the *New York Times* reported, 7.4 million households were engaged in online trading.[24] According to Brad Barber of the University of California at Berkeley, from 1995 through mid-2000, investors opened 12.5 million online brokerage accounts.[25] The busiest year on record was 2013 for both Ameritrade and E*Trade, whose average trades per day were up 24 and 25 percent, respectively, from 2012.[26] But this increased access to the stock markets does not mean people have gained access to the game as the professionals have been playing it.

Studies of this new population of do-it-yourself traders invariably show that increased access to trading tools and market data creates the illusion of market competency and encourages poor decision-making. Even before net access, do-it-yourself investors tended to make poorer investment decisions than those who used financial advisors or, best of all, invested in entirely unmanaged "index" funds. The main reason for DIY investors' poor results? Amateurs trade too much. Meanwhile, online trading brokerages—whose profit comes almost solely from commissions on trading—have a stake in getting their users to make as many trades as possible.

As the data shows, the same brokerage houses profit from the same trading activity in the same way they always have, while retail investors' actual percent of profits and trading accuracy goes *down*. That all this is occurring during the Internet revolution hasn't helped these new investors make informed decisions either. On the contrary, when people have access to more data through which to predict the future, their confidence in their predictions increases much more than the accuracy of their predictions.[27] So the more data an online site can provide its traders, the more secure they will feel trading. Likewise, the more control a trader feels over his activities, the more trades he will make and the more conviction he will have about them—and the poorer he will do.[28]

Fully aware of the psychological influence of information access and the illusion of control, online trading brokerages develop advertising campaigns that exploit both of these vulnerabilities. eSignal proclaims, "You'll make more, because you know more."[29] An advertisement for Ameritrade explains that online investing "is about control."[30] The ever-growing assortment of online trading tools for individual investors may create the sense that the game has opened up but offers the individual investor little more than the vicarious thrill of participation. There are lots of buttons, menus, data points, and choices, but most of them make no real difference to the outcome. The trading itself is real, as are the commissions and losses, but the access to equal footing on the playing field is a digital simulation.

DO ALGORITHMS DREAM OF DIGITAL DERIVATIVES?

Professional traders face similar challenges in the digitized environment. Like those in other disrupted industries, many finance workers have been replaced by networks and computers.

Human stockbrokers, in addition to providing access to markets, used to be responsible for giving clients the best information on a stock or sector, as well as advice and numbers on allocations and future earnings. Trades then went through the specialists—designated market makers—who owned a pool of a particular stock, which they used to fill orders when there was no ready counterparty. They were required to serve as buyers of last resort, preventing a stock from crashing unnecessarily due to a temporary lack of liquidity. Yes, they made money doing this, exploiting their privileged position on the trading floor to buy low and sell high. But their activity reduced volatility and kept markets more consistently liquid and orderly.

Thanks to the net, customers can now access markets—or at least virtual trading desks—directly. The digitized marketplace doesn't require brokers or specialists to function. Trades are executed via "straight through" processes that connect buyers and sellers from around the world. Real-time quotes from electronic trading desks make pricing more transparent; the decentralized nature of the networked exchanges anonymizes deal making and reduces favoritism; and digital record keeping increases accountability and archives any nefarious practices for future audit and prosecution. With multiple and competing exchanges instead of just one specialist, traders also get tighter spreads between the ask and bid, which means less money is leaked out of each trade.

But the elimination of human specialists also eliminates the buyers of last resort and the dampening of volatility they provided. Into the power vacuum and hungry for this increased volatility come the high-frequency traders (HFTs) and algorithms—computer programs that look to profit exclusively through the exploitation of temporary, even microsecond-long, imbalances in trading.

Sure, in many cases, having an algorithm that scans for a lack of volume of a particular security on an exchange somewhere is a good thing. The high-frequency trading algorithm may charge a few pennies extra for having found the stock someone wanted, but the specialist charged a few pennies for his services, too. The problem is that, unlike the specialists who were obligated to reduce volatility in the shares they serviced, HFTs *like* volatility. For instance, one typical HFT strategy is to provide liquidity in a particular stock until the market shows signs of instability. The algorithm then suddenly withdraws all its bids and offers, leading to an immediate dearth of demand and a precipitous price drop.

The smart algorithm, knowing it can make this happen, has already bet against the stock with derivative options. When the other algorithms realize what's happening, they freeze up, too, leading to a "flash crash." The stock goes down, but for no real-world reason. It's just collateral damage from the game itself.[31]

Another common algorithm strategy is to flood the quote and order systems with fake trades—orders of intent but not full executions—to convince human traders (or other algorithms) that the market is moving in a particular direction. More than 90 percent of all quotes are fake gestures of this sort, generated by computers.[32]

Algorithms run on ultrafast computers connected as physically close to the stock exchange computers as possible. This gives them a processing and latency advantage over their peers and any remaining human traders. A well-located algorithm can take action between the moment a more distant human or computer makes a trade and the moment that trade is fully executed.[33] This is essentially "front-running" other people's purchases—buying the shares they intend to buy and then selling them to those same traders at a profit. Although it may cost the trader only a few pennies extra, those pennies add up when algorithms perform these routines millions of times a day, extracting real value from the market.

To be clear, the algorithms are providing no service. Any liquidity they might create is more than compensated for by the liquidity they take away when they're seeking to generate volatility or panic selling. They disadvantage not only human brokers but also the individual investors that a digital stock market was supposed to empower. Algorithmic trading doesn't happen on a laptop connected to the net by Wi-Fi. It requires the kind of hardware, connectivity, and real-estate location that only the wealthiest, most established firms can afford. It may be disruptive to trading, but it only enhances the advantages of the traditional players—or at least the firms they worked for before they were replaced by machines.

New exchanges are emerging to counteract some of these trends. Ironically, they evade algorithms by slowing down the speed at which they communicate their trades. As we've seen, algorithms exploit traders by intercepting their communications as they make their way from one trading desk to another. If a trader is selling a thousand shares of a stock, the order typically goes to a few different virtual trading desks. Because these desks are located in different places, the order may get to one desk a half second before it gets to the second one, or even a whole second before it reaches the third. An ultrafast computer can see what's happening on the first trading desk, then race ahead to the second and third desks and front-run the trade before the original order reaches them. To counteract this, the new anti-algorithmic exchanges calculate how long it will take orders to reach the furthest exchanges, then send orders out in a time-delayed fashion. They send the order to the furthest exchange first and to the closest exchange last. This way the order reaches all the exchanges at the same moment. Even if the algorithm witnesses the first order, it can't front-run the others, because they've all been executed.[34]

And for right now, anyway, that's what constitutes a winning strategy on Wall Street. It's a game being played between algorithms exploiting the trading protocols. It has nothing to do with providing capital to growing companies, and everything to do with extracting value from the investment economy by undermining the very premise of open markets. It is gaming the system.

Stock is intended to be an instrument through which entrepreneurs can raise capital for new businesses or expansions. In exchange for cash, the investor gets a piece, or share, of the company. Once that initial transaction has been completed, however, the rest of that share's journey is inconsequential. Any increase in share value goes to the trader, not the company. The only way the company gets more capital is by selling or issuing more shares.

Still, the value of a share of stock is related to the fortunes of the company. If a company grows, shareholders own pieces of a bigger pie; if the company makes sufficient profits, these are paid up to shareholders as dividends. That's why investors traditionally make decisions based on research about a company, its management, and the business conditions surrounding it.

In contrast to investors, who are looking to grow money over time by assessing the true value of companies, traders seek to profit from the changing prices of stocks and bonds. The underlying worth of a company doesn't really matter. The trader is looking at ebbs and flows, trend lines and moving averages, bubbles and crashes. For the trader, the massive amounts of data and processing capabilities unleashed by digital technology are important only insofar as they offer new ways of strategizing moves in the game.

Digital publishers from Bloomberg to Yahoo Finance are more than happy to satisfy the traders' insatiable appetite for charting and data visualization. Acting a bit like algorithms themselves, traders employ stochastics and momentum oscillators to bet on volatility itself. Of course, their trades are effortlessly preempted by the real algorithmic traders—the algorithms themselves. Those algorithms, in turn, battle other algorithms, all competing to strategize on top of each other's strategies, in successive layers of derivative transactions, each one more abstracted from the last.

Those who play in this space, from individual technical traders to the operators of algorithmic programs, feel they have gotten into the very core of the game—the rule writing itself. In actuality, they may be dominating the landscape of trading, but they are operating on a plane far removed from the companies and investors whose shares they are leveraging. The tremendous volume of activity they generate only disconnects the marketplace further—and extracts more value—from the commerce they were originally created to fuel.

The sheer volume of derivative finance dwarfs that of the real markets. In 2013, the value of the derivative market was estimated at approaching $710 trillion[35]—that's almost ten times the size of the nonderivative global economy[36] and forty-seven times the size of the U.S. stock market.[37] Less conservative estimates put the derivatives market at $1.2 quadrillion, or more than twenty times the world economy.[38] It's only fitting that in 2013, a derivatives exchange called Intercontinental Exchange ended up purchasing the NYSE itself.[39] The stock market—already an abstraction of commerce—was swallowed by its own abstraction.

As markets are increasingly driven by all this virtual gamesmanship, they become more volatile and harder to read in terms of any fundamentals. Minor swings of companies and sectors lead to bubbles and crashes of unprecedented speed and magnitude. The sheer turbulence of all this digital trading creates something of a weather system all its own.

Indeed, the more that algorithms dominate the marketplace, the more the market begins to take on the properties of a dynamic system. It's no longer a marketplace driven directly by supply and demand, business conditions, or commodity prices. Rather, prices, flows, and volatility are determined by the trading going on among all the algorithms. Each algorithm is a feedback loop, taking an action, observing the resulting conditions, and taking another action after that. Again, and again, and again. It's an iterative process, in which the algorithms adjust themselves and their activity on every loop, responding less to the news on the ground than to one another.

Such systems go out of control because the feedback of their own activity has become louder than the original signal. It's like when a performer puts a microphone too close to an amplified speaker. It picks up its own feedback, sends it to the speaker, picks it up again, and sends it through again, ad infinitum. The resulting *screech* is equivalent to the sudden market spike or flash crash created by algorithms iterating their own feedback.

Traditional market players scratch their heads at these outlier events because they can't be explained in terms of trading activity between humans. What made that bubble burst? Was it market sentiment, a piece of news, or something being overbought? None of the usual suspects indicated trouble. That's why it has become popular to label these gaps in rationality "black swans"—as if they are utterly unpredictable anomalies.

In fact, they are entirely predictable. We might not know exactly when these extreme events are coming, but we know they will, because that's the way nonlinear systems express themselves. We are not witnessing momentary crises in the capitalization of business; we are watching a high-stakes video game among the nonhuman players of the wealthiest investment houses. At best, we humans are carried along for the ride.

These boom-and-bust cycles—whether in the microscopic moment of a single hijacked trade or in the macromoment of a major market crash—are the extractive function of the stock market at work. As in a self-similar fractal, the same process occurs on all levels simultaneously—bubbles within bubbles. Every time we humans bid on a stock, we trigger a computerized auction among algorithmic sellers that creates a miniature bubble, just for us—forcing us to buy at the "top" of that micromoment's trading. We humans also lose in the long run, hanging tight to our portfolios as algorithms sell on market highs, dissolve a few decades of our investing, and then buy again at the bottom, after taxpayers have "bailed out" the financial institutions pretending to have been decapitalized in the process.

These chaotic systems may exhibit emergent behaviors and even predictable patterns, but they have almost nothing to do with the underlying landscape from which they spring forth. This is why traditional business analysts, economists, and central bankers alike are at such a loss to comprehend markets in traditional terms. As former World Bank Senior Economist Herman Daly puts it: "Just as in physics, so in economics: the classical theories do not work well in regions close to limits."[40]

The process of capitalization has been accelerated into something other than itself. Instead of integrating the marketplace, digital technology generates derivative systems that extract value for their operators through sheer churn. It's synthetic growth.

Only a true digital native could understand this as a way of doing business.

INVESTMENT GAMIFIED: THE STARTUP

When investing gets so separated from real economic activity, finding funding for a company—without falling into the growth trap—is hard. Entrepreneurs must play the same abstracted game as investors but from the other side of the board.

One of the smartest technologists I know, a young woman from the West Coast I'll call Ruby, decided to launch a company on a whim. She was not interested in making money or even promoting a new technology; she wanted to test her theories about how the ebbs and flows of the startup market worked and whether she could win at the game by getting herself acquired.

So Ruby did exhaustive research on emerging interests and keywords in the technology and business press, as well as conference topics and TED subjects. What were venture capitalists getting interested in? Moreover, what sorts of technical skills would be valuable to those industries? For instance, if she concluded that big data was in ascendance, then she would not only launch a startup related to big data but also make sure she created competencies that big data firms required, such as data visualization or factor analysis. This way, even if her company's primary offering failed, it would still be valuable as an acquisition—for either its skills or its talent, which would be in high demand if her bet on the growing sector proved correct.

She ultimately chose geolocation services as the growing field. She assembled teams to build a few apps that depended on geolocation—less because the apps themselves were so terrific (though she wouldn't complain if one became a hit) than because of the capabilities those apps could offer to potential acquirers. Working on them also forced her team to develop marketable competencies as well as a handful of patentable solutions in a growing field with many problems to solve. The company was purchased, for a whole lot, by a much larger technology player looking to incorporate geolocation into its software and platforms. The employees, founder, and investors who believed in her are now all wealthy people.

Ruby is not cynical; she is a hacker by nature, and merely gamed a system that she knows is already a game. She reverse engineered a startup based on market conditions, industry trends, and nascent investor fads. I asked her if she could do it again—with me as a partner or investor this time. She shook her head. "I'm glad I did it, but it was kind of boring," she replied. "Besides, we're at the wrong moment in the cycle right now. Maybe next year. If you need funding for anything in the meantime, though, just let me know."

Her success may be unique in that she did it as a fun experiment, but Ruby's approach of retrofitting a company to the startup market is all too common. The smartest hackers understand that their skill at hacking technology may be less important than their skill at hacking the digital marketplace. To them, it's all just code—and even if it's not, it's more like code every day. The economy is less a place to create value than a system to game. Hell, everyone in finance and banking is already gaming the system, extracting money from what used to be the simple capitalization of business ventures. Why not create business ventures that game the gamers at their own game?

Besides, most young technology entrepreneurs who come out of college with an idea they truly believe in quickly learn that getting capitalized means sacrificing whatever vision they came in with to the priorities of the investors' game. It's a slow, disillusioning process. Most fail, but failure usually reinforces the need to submit to the game more fully next time. Like devotees of a quack healer, they decide that their poor results must be their own fault and not that of the crazy system to which they have committed themselves and their futures.

The startup game hasn't always characterized Web development. Back in the early days of the net, hackers would create companies almost accidentally. This was the slacker era of the 1980s and early 1990s, when software wasn't particularly valued and life wasn't particularly expensive. Developers could gather in someone's garage and live on pizza and soda as they wrote software and games or prototyped devices and hardware.[41] These were generally ideas that could not get investment capital because they depended on consumer demand that didn't exist yet.

Two kids and a decent computer didn't really need capital to build the next big thing. And while their inventions changed the culture, the companies created by Steve Jobs, Steve Case, Bill Gates, and Mitch Kapor made millionaires out of their founders and first employees. Sure, there were a few friends and industry insiders who had thrown in a little seed money and reaped real rewards, but the process was opaque to the vast majority of the investment community, who were generally shut out of all this until shares became public. By then, many of the companies had peaked, anyway.

As in any pyramid scheme, the real money gets made by those who get in early. So existing venture capitalists, as well as scores of freshly minted ones, came on the scene. This was the late 1990s, when *Wired* said we were in the "long boom," and the Internet development landscape had taken on the quality of a second California Gold Rush. Finding an "angel" with ready cash was easier than finding a kid who knew how to mark up a Web page.

Over the next decade, a basic playbook was established for how a startup gets to IPO or acquisition. Get an idea in college, find a programmer in the same dorm, build a prototype, write a business plan, present it at a conference, do an "angel round," hire a couple more programmers to get to "minimum viable product," raise a "Series A" round of investment, launch on the Web or App Store, achieve or manufacture huge numbers, write a new business plan with some scalable vision, raise a "Series B" round (if you absolutely need more funding), then get acquired or do an IPO. Terms such as *angel round* and *Series A* are now as common in programmer vocabulary as *client* and *server.* And, as young college-dropout CEOs quickly realize, this business vocabulary is more important than coding languages to their success in the startup game.

At one well-meaning Southern California fitness app startup I visited regularly, the young founders held weekly meetings at which the chief technology officer would educate his engineers on different aspects of the development process. But as time went on, he grew less likely to lecture on programming biofeedback interfaces than on business strategy. It was as if he had cracked a new sort of code. He spoke of scalable solutions, long-term contracts, and high switching costs—steps they could take to ensure "defensible outcomes" and achieve a "platform monopoly."

He had fully accepted the startup playbook's emphasis on massive growth above all else and was now turning his tremendous capacity for programming toward that singular, highly limited ambition. The product was less important now than the prize. He and his partner were not in a position to entertain a true disruption of the marketplace, anyway. They had won one of those pitch-your-idea contests by coming up with an idea literally overnight. The venture capital flowed in days later, and these kids were—like so many other young entrepreneurs who accept tens of millions of dollars up front—obligated to build a company worth a billion dollars.

That's the big conundrum facing developers today, and the reason we see so little technological development that goes against the dogma of platform monopolies and other "defensible" outcomes. As we saw with Uber, it's not enough for an app to support a sustainable business; it has to have a path to owning its entire marketplace, presumably forever, with a means to take over still others. Otherwise it can't ever justify the venture capital it has accepted.

Early-stage technology investors aren't looking merely to be paid back with a bit of interest; they want their winning picks to be multiplied *hundreds* of times. That's because they know that only one out of a hundred may end up a winner, and that winner must offset the scores of losers. The earlier and correspondingly more risky an investment, the bigger the required upside. Angel investors, generally the first ones in, may spread out a million dollars over hundreds of different startups—a shotgun-spray approach to high-stakes, low-probability investing. The winning investment must end up paying back at least a million dollars. But to do that, the company has to be worth a whole lot more than that.

Let's say five angels each put $10,000 into a company in return for 5 percent ownership. That's a $50,000 total investment. But the founders want to keep half the company for themselves and their employees, so that means it must begin with a valuation of $100,000. For the angels to earn one hundred times their money, the company must get to a valuation of $10 million before executing an "exit" through which the investors can sell their shares.

Of course, most companies can't exit at that point, anyway. They barely have a product or market yet. The $10 million valuation is based on how promising the company looks to the press, analysts, and the next round of investors. That round, the "Series A," is where bigger venture funds come in. These investors may put in several million dollars, this time according to a company valuation of $10 million to $50 million. The investors at this stage make a similar calculation, spreading their tens or hundreds of millions of dollars across a wide range of startups. They may hope to limit their downside by developing a "thesis" about the kinds of companies they think will succeed, such as "social smartphone apps," "gamified sharing economy," or "health, medical, and fitness entertainment."

But in the end, venture capitalists invest with low expectations of any single company surviving. Their business model is still based on the rare big winner offsetting a dozen or more losers. If they invest $10 million in a company at a valuation of $50 million, then they need that company to become worth *half a billion dollars* in order to see a $100 million return. Even at that, such a return is hardly capable of offsetting the losses throughout the rest of the portfolio. A "win" of that size is the minimum they can justify. In the gamified lingo of the venture capital firms, they can't settle for a "single" or a "double" but must push for a "home run."

That's why those firms usually demand a seat on the board of directors, from which they can steer company policy away from moderately successful outcomes and toward winner-takes-all conclusions. Most venture capitalists would rather drive a company into the ground than let it settle in as a sustainable operation. As long as there is a chance, however small, for a company to become a billion-dollar supersuccess, the investor would rather push on. This means abandoning even surefire profit models if they aren't going to generate the outlandish returns required by the venture capitalist's overall portfolio strategy. He or she would prefer to let the company die, squeezing out every possible megawin, than let it carry on as a moderately successful enterprise. Without a major exit through acquisition or IPO, it is worthless on the level that venture capitalists are playing the game.

I've sat in on more than one board meeting, watching as investors teach their young company founders about the realities of the startup landscape and why they have to shoot for the stars. Every company must become the universal solution in its vertical—or more. *You are not just a personal health app; you are the platform through which all health apps will be executed! You are not just a game; you are a gamification operating system and social network!*

A company is not allowed to reach total dominance incrementally. Venture capital is not patient money. If a company doesn't hit in two or three years, it's considered cold and may as well not exist at all. So instead of developing a long-term strategy, companies make quarterly and semiannual plans—each with its own fantasy megaexit. If they don't get immediate traction, they are supposed to "pivot" to another option next quarter, again and again, until they either hit the jackpot or spend all the money that's been invested. The speed with which the startup burns through its pile of investor cash is called "burn rate." The more employees a company takes on, the more attractive it is as an "acqui-hire," but the faster the burn rate and the shorter the "runway" it has to reach liftoff. If it burns through the cash, it either dies or does another round of fundraising—something the original investors approve of only if the valuation of the company is raised so high that the paper value of their initial investment actually goes up. The higher the valuation, the bigger the subsequent home run must be.

Throughout this whole period, the company must be careful not to show any revenue. That's right: if a company starts taking in money, then it can be judged on its profits or losses. It starts looking like a business rather than a business *plan*. That can kill the all-important valuation right there. Instead, companies need to maintain pie-in-the-sky outlooks about *potential* revenue, once the universal promise of their technology or platform becomes clear to the world—or at least to a buyer like Facebook or Google.

Remember, the venture capitalists got burned by a lot of these kids early on—in either the dot-com bust or the first social media startup bubble. They did not understand digital technology or networking—only that there was something going on that would be big someday. They were forced to nod and accept whatever young tech mavericks like Steve Jobs or

Sean Parker were telling them. Not so anymore. Today, it's the funders who call the shots and the young developers who hang on their every word. These kids, who may have been in a college classroom just a few weeks earlier, now treat their investors with the same reverence they once bestowed upon their professors. Actually, more.

So they accept the hypergrowth logic of the startup economy as if it really were the religion of technology development. They listen to their new mentors and accept their teachings as gifts of wisdom. *These folks already gave me millions of dollars; of course they have my best interests at heart.* After all, these are young and impressionable developers. At age nineteen or twenty, the prefrontal cortex isn't even fully developed yet.[42] That's the part of the brain responsible for decision making and impulse control. These are the years when one's ability to weigh priorities against one another is developed. The founders' original desires for a realistic, if limited, success are quickly replaced by venture capital's requirement for a home run. Before long, they have forgotten whatever social need they left college to serve and have convinced themselves that absolute market domination is the only possible way forward. As one young entrepreneur explained to me after his second board meeting, "I get it now. A win is total, or it's nothing."

His investors taught him what my friend Ruby figured out on her own: that creating a company for acquisition or IPO is different from building a profitable enterprise; it's about building a *sellable* enterprise. Startups are not trying to earn revenue (which is a liability); they are setting themselves up to win more capital. They are not part of the real economy or even the real world but part of the process through which working assets are converted into new stockpiles of dead ones. That's all they have really accomplished with whatever digital fad they've foisted onto the market or sold to yesterday's tech winners. They thought they were engineering a new technology, when they were actually engineering a reallocation of capital.

That's why digital entrepreneurs who do win often end up becoming the next generation of venture capitalists. Everyone from Marc Andreessen

(Netscape) to Sean Parker (Napster) to Peter Thiel (PayPal) to Jack Dorsey (Twitter) now runs venture funds of his own. Facebook and Google, once startups themselves, now acquire more businesses than they incubate internally. With each new generation, firms and investors leverage the startup economy more deliberately, or even cynically. After all, a win is a win.

Take OMGPop, a gaming Web site startup that won a spot in the Y Combinator incubator to build social games. It soon enjoyed moderate success with a Facebook game but then couldn't seem to get any traction. With good advice from its venture-savvy mentors—all former startup founders themselves—the company pivoted from one sector to another, looking for a sweet spot. It picked up another cohort of mentors, including the famed startup studio Betaworks, who helped steer the company toward a trending yet underserved market segment: mobile social gaming. The company came up with a free social gaming app called Draw Something, in which users draw little pictures for one another. In just a week, the game proved a hit, so the developers began to create add-on utilities and features that the most ardent users were willing to pay for.

One month after its iPhone/Android launch, the game had earned fifty million downloads and fifteen million active daily users.[43] Without any real sense of how durable this new customer base might be, OMGPop put itself up for sale. It received an acquisition offer from Zynga, the social gaming company responsible for an earlier megahit, the Facebook-based virtual farmer game, FarmVille.

Zynga, for its part, had managed to earn about $30 million in 2011 selling in-game virtual goods to players for their farms. Its success captured the attention of investors, who were desperate for a way to bet on social media. The company figured it was a good enough proxy for the whole sector and rushed to IPO in December of that year, ahead of Facebook and Twitter. By the end of its first day of trading, Zynga was worth about $7 billion.[44] But FarmVille eventually began to tank, and the company had no compelling follow-up game to replace it. Revenue tanked. Zynga proved better at gaming the market than at making games. Making matters worse, once Facebook had its own IPO, people interested in social media companies sold their Zynga stock to buy Facebook's, further eroding its share price.[45] Zynga had to do something with its capital before it was gone. Besides, social networking was moving to smartphones, and gaming had to follow. Draw Something had to be purchased or cloned—but simply copying the other company's product might make Zynga look even less innovative in the long run.

So, in March 2012, Zynga went and bought OMGPop for its Draw Something franchise and the promise of a mobile future. It paid over $200 million for the company,[46] a little more than ten dollars per active user. Almost immediately, however, the user base of Draw Something began to decline—from fifteen million to ten million daily users the very next month,[47] and down from there. Oops. Draw Something was less the gateway to a new population of users than a short-lived fad. Zynga thought it was buying a ready-made pivot for itself to a new market, but it had really purchased a single mobile game product, itself retrofitted to a market moment. And it did so at the absolute top.

For weeks after that, no one knew quite what to say about all this. The founders of OMGPop were good, hard workers. They weren't pulling an intentional con, even if their company had been used by one group of billionaires to extract a few hundred million from a different group of billionaires. Yet when I went to a Betaworks event that month, OMGPop's founders and advisors were all there with strange grins on their faces. There was a palpable sense of giddiness in the air. They sold the company on the exact day it had the most users it ever would. It was a perfect win. A year later, the OMGPop division of Zynga shut down.[48]

Luckily, OMGPop was never intended to save the world, so its founders got what they were after. They wanted to have fun, but their ultimate intention was to have a spectacular exit. That's why they welcomed the intervention of investors who could coach them to hit the home run. To those who see their startup as having some enduring purpose beyond a market hack, such calls to pivot are less welcome. But they soon learn they are powerless against venture capitalists who insist that there's only one objective and that any goal beyond IPO or acquisition is a distraction. By this point, most founders have taken on far too much investment to refuse.

For some developers, this is the moment when all the venture capital they managed to raise during countless roadshows and pitches begins to feel less like capital than a curse. Slowly, they come to realize that their original vision for serving people with a new technology has been lost amid a series of compromising pivots toward an outlandishly improbable megahit. Succeeding in the new world of platform monopolies doesn't even leave room for whatever the company was first created to do. Gone is the technology company that meant for users to be able to create value for themselves or for other developers to take advantage of some terrific new feat of engineering. These openings for distributed prosperity are now recognized as obstacles to market dominance and sealed shut like potential leaks in a battleship.

For most, such misgivings occur too late, only after the deals are signed, the checks are cashed, and the founders' fiduciary responsibility to shareholders trumps their dedication to their vision. In many cases, the financial payoff offsets the personal pain. Becoming a multimillionaire in one's twenties can be a powerful positive reinforcement for playing by the venture capitalists' rules. But for a few, the compromises become incentives to look for alternatives—for an even more fundamental hack: rather than retrofitting one's company to the existing venture funding model, how about using a company to change it?

In their own run-up to IPO, Google's founders still had enough gumption to challenge Wall Street's power brokers. They insisted on pricing their shares through a bottom-up auction that mirrored the way their search engine came up with results. Traditionally, companies hoping to list on the stock exchange hire investment bankers to pitch them to investors and to calculate the highest valuation and share price that those investors will pay. In return, the bankers get a significant piece of the whole deal—usually around 7 percent—as well as the bragging rights for having been chosen to execute it.

Google figured it had enough name recognition and a clear enough value proposition for its potential investors. Plus, it wanted to demonstrate a more democratic style by offering the very first shares to the general public instead of just to the insiders at Goldman Sachs and Morgan Stanley. They did a Dutch auction, through which anyone with a brokerage account could bid on any amount of shares, at any price. After the bidding, the highest price at which every available share could be sold became the price for all the shares—in this case, $85 per share, or a total of $23 billion. [49, 50]

Investment bankers called the auction a "disaster" and claimed they could have gotten Google a much higher price had they been allowed to offer the shares to their usual clients in the traditional, closed fashion.[51] Then those clients could have sold shares to the hungry public on the open exchange and reaped even more. But the only disaster was for investment banking itself. Google had not only bypassed the bankers but also given the general public first crack at its shares. This was an Internet company at the height of its ability to "do no evil." It took the short-term gains away from elite investors who formerly enjoyed exclusive access to IPO shares, and gave a sense of privilege and ownership to the millions of Google users who were actually responsible for the search engine's rise from dorm-room experiment to technology giant.

Unfortunately, it didn't set a trend. Facebook, Twitter, and every other major subsequent tech IPO went the traditional route, letting investment bankers handle their sales to the public markets. Sure, the IPO prices end up being unsupportable, at least in the short term, but that doesn't matter to the venture capitalists, who are less interested in the future of the company than they are in getting the heck out of their investments. The IPO is their exit, not their entrance. Investors and venture capitalists are not about to let the companies in their stables blow 20 or 30 percent of potential opening-day market cap on some idealistic hacker ethos. From what I've learned in the conversations I've had with the founders of some of these companies, they never had a real choice in the matter.

VENTURELESS CAPITAL: THE PATIENCE OF CROWDS

Still, those who have lived through a painful but profitable soul-sucking home run are the best equipped to do it differently the next time out. My good friend Scott Heiferman made millions on his first company, a Web advertising technology firm called i-traffic that he sold to Agency.com at the height of the dot-com boom. But he felt pretty empty afterward and went on a multiyear walkabout that included a stint working at McDonald's, as well as posting a photo a day on a site he created called Fotolog (which has since garnered over a billion photos from thirty million users, mostly in South America).

As if to counteract some of the Internet's ills, Heiferman founded a new company he called Meetup—a simple Web site through which people can convene meetings in the real world, about anything they're interested in. There are Meetups for pug owners of Memphis, knitters of West Boca Raton, and furries of Houston. Meetups gained traction during the Howard Dean presidential campaign of 2000, when they were used as a primary organizational tool, and they have remained popular with organizers of all kinds ever since.

Heiferman took in some venture capital—but only as much as he needed, and mostly from people he knew would be patient. Today, the company is profitable; it funds its operations by collecting small fees from thousands of group organizers and can do so indefinitely. It's just not the path to a home run. Heiferman regularly fields offers from brands wanting to pay the company to sponsor various Meetups, send targeted ads, or give conveners samples of products; the company is sitting on a gold mine of consumer data, too. But Heiferman rejects them all because he aims for Meetup to be a civic platform rather than a platform monopoly. "We're not trying to vertically/horizontally integrate or get into new businesses or invent self-driving space elevators. We know what business we want to be in, it's a big opportunity, and we don't see ourselves as empire-building imperialists."[52]

Most of his investors are wealthy enough and intrigued enough with his mission to let Heiferman keep Meetup on the slow and steady path. As for those who aren't, Heiferman says, "I'm telling investors the truth: we want to generate good profits and get to a point of paying dividends for many, many years. That's not their business model, but that's not my problem." Still, he is exploring ways for impatient investors to sell their shares to others who will be more content with a long-term dividend generator.

Don't mistake Heiferman for an enemy of the market. He's not running a nonprofit, and he believes his company's mission is to accomplish what "government could or should do but won't do because government is broken and they couldn't assemble the talent and energy to do it. This talent is very valuable in the marketplace." And only a market solution can solve the social ills Heiferman means to address.

Rejecting the rules of late-stage tech-bubble venture-capital madness is a better, more resilient, and durable approach to business in a digital landscape. Who better to affirm this than one of the digital industry's most trusted news and analysis sources, PandoDaily. Generally accepted as the "site of record for Silicon Valley,"* PandoDaily raised just over $4 million in capital, granting no single investor more than 8 percent of the company and giving up no board seats to venture capitalists. If the investors had control, they would push the publication to grow faster and bulk up its staff. They would force it to use its capital to hire away name writers from other business publications, both for the appearance of dominance and to make the company more attractive as a potential acqui-hire. This hiring would, in turn, increase the company's burn rate, forcing a new round of investment, wresting more control of the publication from its founders, and shedding more of their personal mission.

But Pando's founder, tech writer Sarah Lacy, had seen all that before—many times—and wanted to steer PandoDaily on a longer, ultimately more profitable path. "Pando could raise more money and spend double

---

* Midas List venture capitalists in 2013 did a study in which Pando was determined to be one of the five most trusted media brands for venture capitalists, along with the *New York Times*, the *Wall Street Journal*, *Fortune*, and TechCrunch.

our burn rate and I'm not convinced we'd get to our goals any faster," she explained in a letter to readers. "What we would risk doing is building the company at an unsustainable cost structure, give up our lock on control, and one day be forced to pivot to a tech platform—like so many other venture funding content companies that have come before us."[53] She's not even certain there would be a home run waiting for them at the end of that path. "I'm not sure how much this venture capital fervor for content will continue," she admits. In other words, if the sort of company she is creating goes out of style by the time she's ready for an acquisition or IPO, then she will have compromised her company's value and integrity for nothing. Even the *odds* are against selling out.

Lacy is content to become a big, old-style content company instead of a platform of content platforms, or whatever the latest venture capital thesis would have her do. And she's okay with this "even if it means we all never sell a single share." In the current climate, that's taken as a radical statement. It shouldn't be. PandoDaily, Meetup, and other mission-driven companies like them see the value in owning a real company instead of selling a fake one. Their second- or third-generation approach to the venture capital conundrum actually looks a whole lot more like old-school capitalization of business than today's absurdly abstracted startup game.

"In the Internet industry, you're basically a custodian of your own idea for maybe three to five years and then you're supposed to sell. That's insanity,"[54] Kickstarter cofounder Perry Chen told Fast Company when he was trying to explain his platform's approach to venture funding. He and Yancey Strickler started the now-famous crowdfunding site with $10 million in 2009 but made investors agree up front never to sell their shares. "We hope that we can return some of these funds to shareholders through some kind of profit sharing or dividend," Chen explained, "and that's it." Six years later, in 2015, Strickler still enjoyed enough authority over the direction of the company to turn Kickstarter into a benefit corporation.[55] None of his shareholders objected.

He's offering his investors something that's anathema to conventional thinking: a way of participating in living commerce, a sustainable mission, and a continual flow of dividends. It treats money less like ice than like water.

Besides, if the Kickstarter platform works as planned, there will be a whole lot less need for venture capital at all. Kickstarter, and other crowdfunding sites such as IndieGogo and Quirky, seek to democratize fundraising. They give small businesses and independent creators a way to bypass investment by instead seeking funding in advance from their future customers. It's how a musician like Amanda Palmer funds her tours and albums, Neil Young funded development of his high-fidelity digital music device Pono,[56] and Lawrence Lessig funded his super PAC, Mayday.[57] Individuals have raised a few hundred dollars to produce products from coloring books to news articles. Filmmakers have raised millions to produce movies, and a team of video-game producers raised over $70 million to develop a new platform.[58]

All that's really happening here is that customers are paying in advance for the products they want to buy. In some cases, this earns them a discount on the final product. In others, early backers pay a premium in return for a souvenir, a token of thanks, or some other confirmation of their participation in the project. They get the satisfaction of driving production—of tastemaking and participation—instead of simply pulling from the consumption side of things. They have a share in the greater enterprise.

From a business perspective, the lender has been disintermediated. Instead of pitching their ideas to a banker, creators get to pitch directly to their audiences. (Yes, it helps to have an audience in the first place—but more than a few talents and products have been discovered through these platforms, which let users spread news of worthy campaigns easily on social media.) The point is not whether this is a better vetting process for new art and products. It's that the payoff that venture funders once expected for risking their capital has been removed from the equation.

That's a major reversal of the way industrialism once removed humans from value creation, and a significant shift in the way people—especially the young people who use these platforms—are learning to think about capital.

Investment capitalism is itself predicated on risk taking. The earlier and more speculative an investment, the greater percentage of the winnings that investor can demand. One could argue that risk may not have value in itself, but in an economy where cash is scarce, those who can take risks with money can demand compensation commensurate with the odds. In capitalism, a new thing doesn't happen without someone risking that the product fails to find a market. By allowing entrepreneurs to take cash orders in advance, these platforms are effectively reversing the clock: we now know the results before the product is made. No cash is risked, because the backers don't pay until the total amount sought has been raised. The only risk is that the project is never completed, but the open market seems pretty good at evaluating competence: 98 percent of projects that meet just 60 percent of their funding goals are fully completed. Startups funded by venture capital do about the reverse, with more than 90 percent of fully funded enterprises failing.[59] The same crowdsourcing dynamics that Upwork or 99designs* use to shift risk onto freelancers can also shift risk off the table altogether. The less risk, the less money is owed to the risk taker.

So, used appropriately, the net disintermediates the funder, eliminates the need to abandon ongoing productivity in favor of a quick exit, spares the marketplace from having to pay back investors, keeps cash in circulation instead of being extracted, and gives regular people the opportunity to put their money toward what they want to see happen.

Although crowdfunding platforms may solve many an entrepreneur's dilemma, they don't address the investor's. From the funder's perspective, Kickstarter and its peers count as "investment" only in the grander sense.

---

* A Web site where designers compete to create logos for companies, and only the winning designer gets paid.

The platforms don't let backers reap financial rewards, no matter how well the comics, movies, and products they have funded end up doing in the marketplace.

For example, when the Kickstarter community provided Oculus Rift with $2.4 million to develop an immersive virtual-reality headset for video games, the crowdfunders didn't share in the payoff, or the jubilation, when the company was acquired a year later by Facebook for an astounding $2 billion. Sure, those who paid $250 or more got the VR kit they purchased, but the thousand or so people who gave less than $250 didn't even get an unassembled prototype kit—just commemorative T-shirts and posters, the sorts of premiums they'd get if they were subscribing to a public radio station. As *Guardian* tech writer Steven Poole wrote on the acquisition, "About 10,000 people gave Oculus $2.5 [million] between them. I for one am struggling to think of a good reason why each of them shouldn't get a proportional share of that $2 [billion] sale."[60]

They shouldn't for the same reason that those of us writing for Arianna Huffington didn't get a piece of the Huffington Post's acquisition by AOL: because we weren't in on it. Our labor and funding was crowdsourced but in a one-way fashion. In both cases, an Internet platform allowed for an independent operator to leverage and extract value from the network in order to achieve quite traditional economic goals. This is just digital industrialism—the sort that benefits a very few at the head of the long tail.

A number of new platforms are attempting to go to the next level by giving crowdfunders an opportunity to participate fully as venture capitalists in the projects they support. AngelList, a Web site originally dedicated to helping startups connect with angel investors, offers a feature called "syndicates" through which people can back portfolios managed by notable investors, such as author and advisor Tim Ferriss or Launch founder Jason Calacanis. Subscribers pledge to back each of the leader's investments by a certain amount. If the investment works, they then pay between 5 and 20 percent of profits to the leader and 5 percent to AngelList.[61] While SEC regulations are under revision to lower the barriers to entry, currently most of these platforms are limited to "qualified" investors—people who have at least a million dollars in savings. As of this writing, numerous platforms that give less wealthy people access to the startup investing market are under construction.[62, 63]

While these may be democratic developments, they're also as sure a sign of a bubble as any we've seen. Opening the floodgates to more angel investment doesn't mean there will be a greater number of successful startups; it simply means more money will come in at the very top of the funnel. More startups get funded, but a smaller percentage of them survive and pay off. We're back to net-exacerbated winner-takes-all extremes. Under the pretense of individual empowerment, unsophisticated investors enter a market where good information is even scarcer than it is on Wall Street. True, they have proxied their participation to more experienced investors, but those investors are now competing to find winners in an even more crowded marketplace. The more authority amateur investors gain over the placement of their funds, the more likely they are to be exploited as the lowest levels in new pyramids.

As far as the startups are concerned, the plentitude of all this angel-type funding makes it all the more probable for them to get stuck on the startup treadmill, committed to improbable home runs instead of a sustainable business. If anything, it exacerbates the problem, because now all those angels are in an anonymous, disconnected crowd. Like any other long-distance shareholders, they just want their returns.

Instead of facilitating mass investment in the same old securities and marketplaces through largely the same old extractive intermediaries, networks can promote direct funding among peers. The way to tell the difference—as with any networked commerce—is if the connection is truly lateral. Are you giving money to a platform or to a person? Is that person retaining the value he or she creates, or is that value going to the much larger corporation he or she works for?

Microfinancing platforms seek to foster that peer-to-peer connectivity, and in some measure they succeed. If crowdfunding platforms like Microventures can be thought of as alternative stock markets, then microfinancing sites from Kiva to Lending Club are more like a bond market. People lend money directly to their chosen peers, for a fixed rate of interest.

Kiva functions more like a charity than an investment. Lenders peruse the site for opportunities to help an impoverished person get started in a business. Farmers in Guatemala need two hundred dollars for seeds, seamstresses in Africa want fifty dollars for buttons, and a messenger in Calcutta wants twenty dollars for new bicycle tires. As numerous studies have now shown, microlending works better at growing economies than charity because recipients are under pressure to grow a business and pay it back. This works even better when borrowers are put into small, local groups whose borrowing power is dependent on everyone's credit histories. For example, if a borrower's husband tries to take the money to buy alcohol, the woman's friends will exert tremendous social pressure on him to return the funds.

Unlike World Bank loans, which trickle down through government, if at all, microlending goes directly to entrepreneurs—with no policy strings attached and no further obligation once the loan is paid. The job creation is more organic and long-lasting than when a foreign corporation comes and plops down a factory, and the funds tend to stay within the community instead of being extracted. This is no longer a fringe activity but a primary catalyst for business and employment in these regions.[64, 65] And it's not the sort of activity from which foreign corporations can simply extract value. Once, after I gave a talk about the promise of microlending, a marketing executive from a European cosmetics firm asked whether I thought that the millions of successful loan recipients in Africa might constitute a network through which she could distribute her products. No, I explained, these women are not her future marketers; they are her future competition.

Still, the microfinancing model has served as a proof of concept for less altogether altruistic platforms, such as Lending Club and Prosper Marketplace, which connect borrowers directly with lenders looking for returns. The sites let prospective borrowers list their requests, along with personal information and stories about their experiences and aspirations.

Lenders peruse the listings, then choose the loans they wish to finance. Because they are connected more directly, borrowers pay less than they would to a bank, and lenders receive more than they would in interest on a savings account—usually in the high single digits.

Borrowers defaulted significantly less on these platforms than they did on traditional loans, largely because the psychology of failing to repay another human being is much different than that of owing money to a faceless bank. To magnify this payback effect, lending platforms learned to use faces of real people in their ads and communications.

Unfortunately, the microfinancing industry did so well that it became plagued by the very force it was designed to sidestep: institutional capital. These platforms took on significant investment from the venture capitalists and needed to scale much faster than their individual subscribers could support. So the platforms welcomed the participation of banks and other institutional lenders that could provide volume and justify higher company valuations. As of early 2014, Lending Club and Prosper Marketplace alone facilitated over $5 billion of loans. Lending Club did its IPO in December 2014, at a valuation of about $10 billion.[66]

Although institutional lenders help these platforms justify their valuations, they push out the human users and undermine any peer-to-peer ethos or activity. The banks and credit companies on the platform use algorithms to cherry-pick the best loans for themselves as soon as they are listed (like professional shoppers taking the best clothes from the thrift shops), leaving only the dregs for the regular people visiting the site. These institutions enjoy the benefits of all that peer-to-peer psychology, earning a sense of loyalty usually owed only to other human beings. But slowly, as these sites' human lenders become aware that they no longer have equal access to the best opportunities, they leave, disillusioned. The nascent peer-to-peer lending landscape is discredited before it even has a chance to propagate.

The sweet spot in digital investment is for investors, lenders, and the enterprises in which they want to participate to be able to create value together without surrendering the human connection that elevates the success rate of peer-to-peer business solutions. As we have seen before, this doesn't necessarily mean exploiting digital technology so much as the digital sensibility.

The most promising new structure I've come across so far isn't really new at all but the repurposing of an old one, called the DPO, or direct public offering. It's a legal structure that was most famously retrieved and employed by Ben & Jerry's when it was seeking its first $750,000 of capital from 1,800 ice-cream-loving Vermonters to build a new plant.[67] (It would have been wise to continue on this path instead of reaching for an IPO, losing control of the company to anonymous shareholders, and getting acquired by Unilever.)

The DPO allows small and medium-sized businesses to raise investment capital from any number of accredited and unaccredited investors—as long as they do it within their own state of incorporation. Unlike an IPO, the DPO happens on a state level, where it isn't subjected to the expensive and arduous vetting process; and unlike crowdfunding, a DPO can offer equity and dividends instead of just a payout on exit. Most important, a DPO gives a business a framework through which it can raise money from the people who know it, work for it, and buy from it.[68] It's like a "friends and family" round, through which entrepreneurs can embed their values right in the capitalization structure, as if programming an ongoing operating system. Rather than platform engineers or even well-meaning intellectuals coming up with more equitable business models, people in the trenches—enabled by a more distributed digital sensibility—often figure out the best approaches simply by solving their own problems.

For example, after a decade of growth, Dan Rosenberg and Addie Rose Holland's organic cannery, Real Pickles, needed to expand their operations. They knew that traditional financing could force them to compromise their commitments to treating workers properly, sourcing materials locally, and using organic practices. They wanted to accept funds without diminishing their workers' role in the company—particularly in the distant future.

So they filed for a DPO, which cost them $15,000 in legal fees but let them raise half a million dollars from seventy-seven investors—all from among their existing network of grocers, small farms, longtime customers, and supporters, who were happy to ensure the success of a company they saw as a vital member of their community and, for many, an important business partner.[69] Imagine having a company in which your investors were also your suppliers and customers; each stakeholder has multiple stakes in your success and multiple ways of ensuring it.

Best of all, the looser regulations on DPOs let Rosenberg and Holland tailor the conditions of their offering to their own values. They wanted to be able to resist the pressures for fast growth that they believed had compromised many other originally mission-driven companies. So they created a rule stipulating that investors have no voting rights and that they must wait a minimum of five years before cashing in their shares. When they do, they receive only their original purchase price. Any return on investment would come from a dividend. Meanwhile, part of the new valuation of the company went to giving workers shares in the enterprise, without the same obligation to retain them for five years.[70]

By rejecting the one-size-fits-all logic of industrial investing, businesses restore a modicum of agency to their own operations. Instead of using digital platforms to amplify the expectations and liabilities of traditional venture capital, they can employ a hacker's DIY sensibilities to build locally capitalized companies on their own terms. Investors, meanwhile, gain the ability to support and benefit from network connections that flow in more than one direction at the same time. Instead of extracting value, they exchange it.

FULLY INVESTED—FACTORS BEYOND CAPITAL

We can't survive in a real-time, entirely liquid economy, no matter how much we might want to. Bad things happen that can destroy a person's ability to make a living wage. People get old and find themselves without a community or family to embrace and support them. Even a limited ability to invest capital, grow savings, and retire on those funds is better than nothing. We have all, to some extent, depended on the industrial economy's ability to preserve wealth over time: to lock in earnings and then grow them at least at the rate of inflation.

The new digital landscape reverses a lot of this. Nothing stands still. The shift from physical paper to charged pixels is a great metaphor for how everything is rendered dynamic and fungible by digitality. The facts in the encyclopedia used to just sit there on its pages; they were both unchangeable and undemanding. The facts in Wikipedia need to be actively maintained on servers and then delivered to those who call them up.

Likewise, as we've seen, digitality changes money (over a number of steps) from a physical piece of gold to a charged particle; from an institutionally guaranteed value to a collaboratively maintained ledger. The former might need to be protected by a vault, but the latter needs to be kept alive with electricity. It's always in an energized state. Money is moving off the hard drive and into active RAM. This makes it harder for anyone but the platform monopolist to make money by standing still—and even the platform monopolists are losing their grasp on the economy. Money alone won't buy security, and everyone is going to have to learn how to invest through work, active participation, and assets other than cash.

Not to worry: this isn't all happening so fast. As disappointing as it may be to the revolutionaries among us, the traditional debt-based investment economy is not flipping into a real-time, distributed, peer-to-peer, cryptocurrency marketplace overnight. Those of us with jobs and families and mortgages ignore the investment markets at our own peril. And there are still ways to invest plain old money that capitalize on the current economic transition without overly compromising our potential for a more equitable economic future. I'll briefly touch on some of these strategies now because, believe it or not, there are readers who came for this alone and have been skimming to this point. I've waited until the end of the book as a way of forcing them to absorb a little something about how the digital landscape can promote entirely new value creation and exchange, even though all they really want to know is which ticker symbols they should buy in the meantime.

1. No Growth, No Problem: Invest for Flow

So fine: the key to investing in the emerging digital landscape is to diversify. The Talmud instructed ancient Jews to keep one third of their assets in land, one third in commerce, and one third "at hand."[71] This would mean real estate (your home), speculation (business, stocks and bonds), and cash or currency equivalents like gold. The real estate is a capital investment that grows, the commercial securities should produce dividends (with risk), and the cash is there to be spent or used for emergencies. These are roughly the same as the three main components of Bernard Lietaer's proposed currency: trees, highways, and gold.[72] Trees are the real estate component, only they have built-in growth. Toll-collecting highways provide the business component in a relatively stable way. And gold is the on-hand cash. The interesting thing to note about both trios is that the business component is not understood as the way to achieve capital gains—that's what the property is for. Instead, the focus is on business investing as a productive, real-time asset generator.

That's also an appropriate strategy for investors who understand the way a properly functioning digital economy will diminish the power of standing capital to suck value out of real commerce. At the very least, it will be for those who don't *want* their own security to depend on sucking the value out of real commerce. It's a path to making money that's not dependent on the entire economy continuing to grow. Investing in flow over capital accumulation is as easy as picking a few high-dividend stocks in sectors that are consonant with your own values, or a dividend-focused mutual fund or ETF.

Of course, you have to do a bit of research. Read quarterly reports and interviews with management—not to time your trades (pointless) but to see how CEOs are incentivized and whether the company has share-price targets (bad) or earnings targets (good). What are the company's profits relative to its total asset value, and is that ratio going up or down? These are better indicators of sustainable revenue for you, as an investor, than the metrics touted by the professional traders on business shows that treat the market like fantasy football.

This is not radical or new, and it works in any market environment. Warren Buffett, one of the most conservative and successful investors of all time, invests primarily in dividend-paying stocks and holds them for a very long time. The top five companies in his Berkshire Hathaway fund—Wells Fargo, Coca-Cola, American Express, IBM, and Walmart—may not be the enterprises we want to support, but they all pay high dividends and increase their payouts regularly. These are stocks that literally pay you to buy and hold them. By refusing to sell a stock, you also avoid the taxes, commissions, and other frictional costs of a portfolio in constant turnover. Instead of profiting from the change in the price of the stock, you profit from the constant flow of revenue from the business itself. As Buffett puts it, "Our approach is very much profiting from lack of change rather than change."[73]

You still have to be careful; some companies produce dividends in pretty abstract or destructive ways. You want to be invested in companies that create value through flow of revenues, not the extraction of a fixed resource from the ground or an asset from a community. Other companies just do evil things you won't want to be rooting for. Sure, there's some karmic safety in the fact that buying a share of a cigarette company is not an investment in cigarettes. As we have seen, the stock market is utterly abstracted from whatever businesses those stocks may have financed decades ago. You're buying shares from traders, not from companies. In that sense, a share of a munitions manufacturer is no more morally encumbered than a share of an organic-food supplier. But when you invest for flow, you are much more dependent on the ongoing profits of the company. This means your dividends may depend on sales of the very things you detest and the escalation of activities from smoking to terrorism that fuel those sales.

Numerous filters have been developed to help people invest more responsibly. Mutual-fund companies such as Ariel and Calvert offer modified versions of the S&P index, filtered for liquor, cigarettes, weapons, and others of the most objectionable industries. Many of these funds also check to see whether a corporation offers overtime and maternity leave to workers, supports the communities in which it operates, and has an environmental responsibility policy. But many companies have learned how to meet the requirements of social responsibility filters without really functioning in a socially responsible fashion, and the indexes are filled with the names of companies most ethical investors would hope to avoid.

2. Bounded Investing

To invest in companies that promote your social or even practical goals, it's a whole lot more straightforward if you start closer to home. Many union pension funds, for example, target their investments toward companies that employ the unions' respective members and directly benefit people in the same socioeconomic bracket. The AFL-CIO Housing Investment Trust funds union-constructed affordable housing in an effort not only to reap profits but also to keep wealth circulating within the working class. The trust invested $750 million in New York City to finance the first construction after 9/11, generating 3,500 union jobs and building 14,000 housing units, 87 percent of which were designated as affordable housing.[74] Another billion went to mortgage loans for union members and city employees. Similarly, the union invested in restoring the Gulf Coast after Hurricane Katrina and in addressing multifamily housing shortages in Chicago and Massachusetts. The investments generated jobs, goodwill, and homes, in addition to retirement fund returns for union members.[75]

Individuals can apply the same principles to their own investments by considering how to leverage the impact of their capital to benefit their own lives. It's easiest to see and feel that impact by investing in regional companies and projects, big or small. A large corporation's local activist shareholders can have a disproportionately significant influence on how the company employs, sources, pollutes, and donates. Meanwhile, as distributed technologies allow smaller, more locally connected businesses to leverage people, assets, and trends that larger conglomerates cannot, consider putting your capital right there. Do you want a local bookstore, gym, or Thai restaurant? Invest in one. Yes, business partnerships between local merchants and friends are stickier than the anonymous purchase of stocks through a laptop, but that's also what makes them better. Investors are more directly committed to one another's success and get to see the fruits of their investment in their own lives. The successful local farm you've funded feeds not only your bank account but your family as well. And your neighborhood. And your property value. And the local economy.

Investing locally, in people and companies you know, may sound at first like conflating business and the personal or overweighting just one kind of sector. But it's actually a form of diversification. We do not just buy a stock with a single benefit of growth; we buy into businesses whose operations benefit us and our communities. We don't need them to keep growing; we just need them to keep operating, generating revenue and benefits. This is part of what was originally meant by double bottom line: the returns on investment take multiple forms. If we invest in companies that in turn pay us for our products or services, we end up generating a double stream of revenue. It's not limited so much as self-sustaining.

We might better call this a strategy of *bounded* investment. By identifying a community, interest group, or region to support, we create boundaries for our capital. The money injected ends up circulating through the pool instead of being sucked out by a company foreign to the system. By targeting certain companies within the pool, we can create better customers and providers for others in the bounded network, including ourselves. These boundaries are not elitist; they are what distinguish connected, living businesses from those that merely extract our energy and money and convert it to capital in a distant shareholder's account. Boundaries don't exclude participation so much as allow for the same investment to touch and strengthen dozens of businesses in the same pool.

Bounded regional investing contradicts the corporate strategy of reducing municipalities' authority over their own laws and land. Investing in a local polluter doesn't increase the value of one's home; investing in an organic market might. Moreover, in a networked era, boundaries don't have to be solely geographically defined; they simply have to define a mutually supportive range of businesses. Your target could be the business sector in which you work, such as design services, equipment, and Web sites. Or your pool could be the various constituencies in biodiesel manufacturing, comic-book publishing, or natural health care. As long as there's a network of businesses that support one another, the boundaries make sense.

Unlike traditional shareholding, bounded investment is less dependent on growth than it is on sustainability. That includes the environment in which customers live and work but also the revenue streams of the businesses in the pool. Where an unbounded investment must continue to grow to be worth anything, a bounded investment needs only to promote the health of the others in the network and sustain itself in doing so. Unbounded investing is like air-conditioning one's home with the windows open. Bounded investing permits an accumulation of assets, expertise, relationships, and monetary momentum.

Bounded pools also free up people and organizations to invest in more than one way—to diversify not only in investment targets and measures of return but in the *means* of investment. In a world where capital itself is inaccessible, dysfunctional, and losing its value, we need to invest with things other than our money—if we even have enough to invest in the first place. This means spending our time, our effort, our social capital, and often our sweat. Our work is our investment.

The objective here is to find ways to structure work so that, more than mere employment, it constitutes a form of ownership. We can sing for our supper, but we want to gain some traction as we do. Currently, that means outsourcing the investment of excess salary to the financial services industry. Instead, we need to find ways for labor itself to be understood as an investment. Employees must earn more than cash and a few token shares of company stock. They should own the companies they work for. Their noncash contributions to the enterprise must be valued as much as investors' capital.

Economists have long understood that it takes more than money to create goods and services. Labor, land, and capital—together—have been recognized as the "factors of production" since even before the classical economics of Adam Smith. Some add entrepreneurship as a special category of labor, but it was obvious to all that enterprise requires work and physical resources in addition to seed money. (Labor, land, and capital are analogous to that Talmudic trio of business, real estate, and cash.)

Thanks to the rise of the finance industries, capital has diminished the market value of the other two factors. Money is the only one that counts as investment anymore. Labor and land have been reduced to externalities—rented, disposable commodities. The only way to invest your land in an enterprise is to take out a mortgage and use the cash to buy some stock. The folks contributing their automobiles and driving labor to Uber, or their property and hosting to Airbnb, make less than minimum-wage employees and don't own a piece of the company even though they constitute the infrastructure. Only money talks.

The preeminence of capital—its ability to elect leaders, dictate law, and reward itself—was a predictable outcome of the industrial age. Global expeditions, big machines, and centralized production were much more dependent on capital than the small businesses and local craftspeople that preceded them were.

Moreover, of the three factors of production, only money is unbound. Well, centrally issued debt-based currency is unbound. Capital is unbound, and must be, because it has to be able to grow infinitely in order for interest to continue to be paid up to its issuers. That's why capitalists advocate open markets, free trade, and minimal regulations. They are intolerant of anything that would serve as a boundary to the unfettered movement and growth of capital.

Labor and land, on the other hand, are bound. That's why they are shunned as obstacles to market growth. Communities impede the free market by objecting to the construction of a factory or power plant within their boundaries. "Not in my backyard" is interpreted as a selfish stance, because any local environmental destruction pales in comparison to the wider, universal benefits to the marketplace. Labor unions, for their part, which try to create a bound cohort of workers, are derided as anticompetitive and exclusionary. Labor should instead be sourced from anywhere, at any price the market will bear. By breaking the boundaries around a particular labor group or region of people, business is free to grow.

But no matter how the market might treat them, both people and places are still bound. There's only so much land to develop. People can put in only so many hours before they drop. Despite that, the market attempts to treat these fixed factors as unbound. Why should a factory worry about the land on which it pollutes? There's always some other place to go. There are still plenty of nice trees and clean water in Washington state or Canada. There's always China or Africa for toxic landfill. At least those countries are getting paid to take the waste. There are thousands of miles of unused land on planet Earth; it may as well be infinite. The same for labor: if the people in this town or state won't work for less money, then the people in that state or country will. There are billions of people on the planet. The available labor these people can offer may as well be unbound, too—or so that faulty logic goes.

Digital industrialism seeks to push people and places to their unbound extremes. Amazon Mechanical Turk unbinds laborers from time, place, and one another. Without physicality, the worker has no way of expressing bounds or forging connections. The platform limits side-to-side conversation; there's no chat function. Meanwhile, the net further minimizes the particularity of place—whether it's local merchants providing physical showrooms for customers who later order online, or social networks encouraging us to think of our community as a nonlocal assortment of connected profiles. In such an environment, the value of labor and land becomes, at best, discounted.

Bounded investing is more like running a family business: you're in it forever, because it's the legacy your kids will inherit as well as the place they will likely spend their careers. If your son returns from college unable to find a job, you hire him to work in the company. This lets you spend the same dollars twice: supporting him and getting a worker who knows the business and cares about his inheritance. The business is your home.

Unbounded investing, on the other hand, promotes the don't-poop-where-you-eat philosophy of business, limits the number of times the same dollar ricochets through an economy, and depends on the illusion of an infinitely expanding marketplace. When we adopt even a partially bounded strategy, the money we spend comes back to us time and time again and the investments we make are reflected in the real world in which we live. You want a sustainable economy? Recycle money.

3. Run Your Business as a Commons: Platform Cooperatives

Resistance to digital industrialism may look like communism, but it's better understood as a simple reinstatement of the commons. Most of us became familiar with the commons in its modern form—projects like Wikipedia or Lawrence Lessig's Creative Commons, essentially gift economies in which people participate together for a greater good. In these two cases, the information on Wikipedia and the music, writing, or code assigned a Creative Commons license are assumed to belong to everyone and no one. Nobody owns them, and everyone can share them within the limits set by mutually agreed-upon rules. I may quote Wikipedia in this book as long as I credit the source.

The commons were originally a set of lands in England owned by the Catholic Church and open to local farmers for grazing. There were strict sets of rules about how much land one could graze and how often, which kept the commons capable of sustaining everyone's flocks in a fair fashion. After King Henry VIII rejected the authority of the pope, those common lands became privatized, or "enclosed."[76] Over the next couple of centuries, the myth of the "tragedy of the commons" was employed to bring the remaining common lands under private control.

The premise that economists and early agricultural investors put forth was that if no one owns something, there's no one to protect it. So common lands will supposedly get overgrazed, and common fisheries will run out of fish. Never mind that there's no evidence of either happening in the maintained commons of England. The supposed tragedy became an accepted truth taught in basic economics courses, and as easy to prove as the condition of most public restrooms in the United States. As recently as 1968, ecologist Garrett Hardin convinced the remaining holdouts of the tragedy of the commons by arguing that Darwinian selection favors privatization by the strong, and that if the world's land is not privatized, the results would be "horrifying." In his words, "injustice is preferable to total ruin."[77]

The false assumption is that people are incapable of recognizing the value of their shared resources and then organizing to protect them—and in doing so, create great value for everyone involved. There are hundreds of examples of highly functioning commons around the world today. Some have been around for centuries, others have risen in response to economic and environmental crises, and still others have been inspired by the distributive bias of digital networks. From the seed-sharing commons of India to the Potato Park of Peru, indigenous populations have been maintaining their lands and managing biodiversity through a highly articulated set of rules about sharing and preservation. From informal rationing of parking spaces in Boston to Richard Stallman's General Public License (GPL) for software, new commons are serving to reinstate the value of land and labor, as well as the ability of people to manage them better than markets can.

In the 1990s, Elinor Ostrom, the American political scientist most responsible for reviving serious thought about commoning, studied what specifically makes a commons successful. She concluded that a commons must have an evolving set of rules about access and usage and that it must have a way of punishing transgressions. It must also respect the particular character of the resource being managed and the people who have worked with that resource the longest. Managing a fixed supply of minerals is different from managing a replenishing supply of timber. Finally, size and place matter. It's easier for a town to manage its water supply than for the planet to establish water-sharing rules.[78]

In short, a commons must be bound by people, place, and rules. Contrary to prevailing wisdom, it's not an anything-goes race to the bottom. It is simply a recognition of boundaries and limits. It's pooled, multifaceted investment in pursuit of sustainable production. It is also an affront to the limitless expansion sought by pure capital. If anything, the notion of a commons' becoming "enclosed" by privatization is a misnomer: privatizing a commons breaks the boundaries that protected its land and labor from pure market forces.

For instance, the open-source seed-sharing networks of India promote biodiversity and fertilizer-free practices among farmers who can't afford Western pesticides.[79] They have sustained themselves over many generations by developing and adhering to a complex set of rules about how seed species are preserved, as well as how to mix crops on soil to recycle its nutrients over centuries of growing. Today, they are in battle with corporations claiming patents on these heirloom seeds and indigenous plants. So it's not the seed commons that have been enclosed by the market at all; rather, the many-generations-old boundaries have been penetrated and dissolved by disingenuously argued free-market principles.

On the brighter side, the peer-to-peer nature of so much of our digital activity has led to a revival of commons and commons-inspired activity. The Open Source Ecology project is a collective of engineers, machinists, and designers who have spent the past ten years developing and refining farming and industrial machinery according to open-source principles. The group publishes its schematics freely online, for anyone worldwide to download, build, or improve.[80] These machines cost a fraction of their standard retail counterparts.[81]

OSE believes that corporate profits, exaggerated by abuses in intellectual property law, create an artificial scarcity for such products that is damaging to all but those at the top of the corporate pyramid and that an information commons can correct the imbalance. "The end point of our practical development is Distributive Enterprise—an open, collaborative enterprise that publishes all of its strategic, business, organizational, enterprise information—so that others could learn and thereby truly accelerate innovation by annihilating all forms of competitive waste."[82]

The most ambitious commons-inspired project to date, the Ecuadoran government's "Free, Libre, Open Knowledge" program, or FLOK, seeks to transform the entire nation from its current extractive, oil-based economy to one based on a protected commons of both real and digital resources.

Under these policies (still in development), intellectual property would be considered part of the commons. This would lead to the creation of hyperlocal factories, schools, and labs, freed from the constraints of licensing fees. So the thinking goes, this would then allow companies to operate with greater fairness, efficiency, and sustainability. The FLOK project originates with a specific set of Latin-American socioeconomic concerns. Foremost among these is "biopiracy," the practice of industrial agricultural companies such as Monsanto, which patent organic technology developed over centuries by local and indigenous farmers. Here in the United States, we can find an analogue in the practice of tech giants such as Apple or Google, that rely on the open-source commons for many of their products' architectures, profiting without paying back into the digital commons. In response to these challenges, FLOK proposes the development of peer-production licenses under which only commoners, cooperatives, and nonprofits would enjoy free usage of intellectual property bounded by the commons; corporations would have to pay.[83, 84, 85]

At first glance, the so-called "sharing economy" appears to be based in these commons principles. At least in some superficial way, this is true. We have gone from buying music on records or CDs to downloading MP3 files to simply subscribing to Pandora or Spotify. Owning music—or a car, for that matter—is becoming less important than having *access* to it. This is certainly a step on the path from hoarding to sharing. Except the many sharing platforms and services are not sharing at all but renting. We don't collectively own the vehicles of Zipcar any more than we collectively own Spotify's catalogue of music. And as private companies induce us to become sharers, we contribute our own cars, creativity, and couches to a sharing economy that is more extractive than it is circulatory. Our investments of time, place, and materials are exploited by those who have invested money and actually own the platforms.

Now that we can see it, however, we can also envision the alternative: we join and form businesses that value our real investments of effort, stuff, and community resources. Imagine an Amazon owned by the sellers, an Uber owned by the drivers, or a Facebook owned by the people whose data and attention is being bought and sold. Distributed digital technology makes this not only possible but preferable to the locked-down, overprogrammed, and extractive platform monopolies of today. It's a lot easier and more efficient just to make something work than to make it also eliminate all potential competition in the process. Such applications and companies would be designed not to promote unbounded expansion into all possible horizontals and verticals but simply to sustain its member/worker/owners and their relationships with their customers. By establishing such limits on expectations, these companies are no longer obligated to return hundreds or thousands of times the capital invested. There is no exit, nor should there be.

Could people develop platforms without the venture capital of the current startup scene? Between BitTorrent, Wikipedia, and Bitcoin, we've seen them do it already. And luckily, venture capital has already funded the development of Amazon, Uber, Airbnb, and dozens of other fat platforms that can easily be cloned and deployed in better ways. Yes, there may still be some costs in maintaining servers—but there won't need to be legions of investors behind that basic infrastructure, demanding that a majority of the revenue be extracted from the system.

Instead, we get a new, powerful, and highly accessible version of the kinds of worker-owned collectives that are providing jobs and profits to people in recession-torn Europe and have been for decades. Although their activity is essentially off the radar of Wall Street, they stand as a powerful proof of concept for a digitally empowered, worker-owned business landscape.

For example, in stark contrast to the business bankruptcies and unemployment plaguing Italy as of this writing, in the Emilia Romagna region a full 30 percent of GDP is created by a bounded network of independent, employee-owned firms constituting just 10 percent of the population.[86] Ranging from manufacturing, construction, finance, artisans, even to social service organizations, the firms coalesce into "flexible manufacturing networks" that cooperate on specific projects, quickly pooling labor and resources, then disbanding as necessary. It's essentially a bounded network of worker-owned businesses that source from one another instead of seeking the lowest bidder or attempting to expand into an additional horizontal. Valuing sustainability over growth, the businesses remain small but also agile enough to retool if market conditions change and networked enough to get support if they need to pivot.

Worker-owned cooperatives can even scale. The oft-cited Mondragon Corporation began in the Basque region of Spain in the 1950s when a Jesuit priest and seven of his students formed a worker-owned paraffin heater company. Over the next decades, the cooperative launched more worker-owned businesses, organized through elected councils that coordinate international strategy. Any business may apply for membership in Mondragon, as long as it is willing to become worker owned and governed. United Steel Workers is involved in discussions about launching worker-owned companies as part of Mondragon, which currently employs over a hundred thousand people.[87] Its big-box retailer, Eroski, has managed to keep even Walmart at bay in Spain.[88] But Mondragon's enormity, as well as the scale and scope of its operations, makes it more vulnerable to the problems of traditional corporate conglomerates.

Some of its larger companies have made missteps into ethically suspect offshoring, and a few push for growth as ardently and self-destructively as any traditional multinational corporation.[89] But Mondragon's success does prove that worker-owned cooperatives aren't just for making organic jams; they can scale vertically and horizontally and compete with any shareholder-owned corporation on the block—without sacrificing the value of land or labor to that of capital.

Digital technologies can distribute these cooperative principles even further, giving independent workers and creative freelancers the ability to organize into co-owned networks instead of surrendering their autonomy to platform monopolies. These platform co-ops are the current vanguard of the digital economy. Most are still in the developmental stage, but then again, the Web was still in it infancy when I published my first book about the Internet, and the drive and need for cooperative platforms is more urgent now than the demand for a Web browser was in 1991. And they are beginning to emerge. Whether or not any of these particular experiments survives as a company, they represent a new willingness to fund and run businesses in a distributed fashion.

EthicalBay, a cooperative version of eBay launching in the United Kingdom, will be owned by its producers and consumers. "Being ethical begins with who owns the company," explain the founders.[90] The company will not accept venture funding (not that anyone is offering), which they hope will allow them to value human contributions and the sustainable use of resources over the short-term needs of capital. Fairmondo, an open-source, user- and worker-owned e-commerce platform, has operated in Germany since late 2012. On the consumer end, Fairmondo's interface is similar to that of any major e-commerce site, offering new and used books, apparel, and electronics. However, unlike Amazon, eBay, or even Etsy, Fairmondo allows users to offer items for trade or exchange or to lend them free of charge. It also allows shoppers to filter searches for ecological and fair trade practices.[91] Ownership in the Fairmondo platform is open to all its users.[92] Profits are shared proportionally to a member's holdings, but everyone has the same single vote in the governance of the company,[93] and the highest salary within the company may not exceed three times the lowest.[94] The platform is open source, and the company subscribes to "open innovation" practices; operations are radically transparent, including fully open books subject to member review. A quarter of profits go to fair trade and environmental organizations bounded by the collective's local, personal, and business networks.[95]

In terms of fully decentralized commerce, these platform cooperatives are still just steps along the way to digital distributism. As long as there's a central platform—a Web site or other hub to maintain—there will always be a need for central funding and an emphasis on command and control. At the very least, there will be an ongoing dynamic tension between the provider-owners on the periphery and the manager-facilitators in the middle. As P2P Foundation founder Michel Bauwens puts it, "a truly free P2P logic at the front-end is highly improbable if the back-end is under exclusive control and ownership."[96] Trust and authentication are rendered by the platform, which may not necessarily end in extractive monopoly but will remain biased toward more industrial-style models and behaviors. That's what's threatening Mondragon's integrity and operations today.

The answer, of course—the last piece of the puzzle—would be to replace platforms with protocols. This is where an intrinsically distributed technology like Bitcoin's blockchain comes in. Instead of depending on a central bank or other authority to authenticate a transaction, interested parties maintain authority and ownership of the Bitcoin blockchain themselves. The bank is not a ledger in a vault or a file on a server but a public record on everyone's devices. As blockchain developers are proving, this ability to administrate through protocols instead of platforms can extend from currency transactions to company operations.

For instance, imagine a platform-independent Uber, owned by the drivers who use it. There's no server to maintain, no venture capital to pay back, no new verticals or horizontals in which to expand, no acquisition, and no exit. There are just drivers whose labor and vehicles constitute ownership of the enterprise. One such experiment, La'Zooz, is a blockchain-managed ridesharing app, where the currency (Zooz) is mined through "proof of movement."[97] So instead of supplying and driving cars as underpaid freelancers for Uber or Lyft, drivers are co-owners of a transportation collective organized through distributed protocols.

Could such platform cooperatives catch on? The basic behavior of downloading an app in order to work or rent property has already been anchored in users by Airbnb, Uber, TaskRabbit, Mechanical Turk, and countless others. Using a blockchain is just a small step further, compared to the original leap into digital labor and exchange. It is the disintermediation that all these supposedly disruptive platforms were promising in the first place. Of course, just because we have the capability to employ protocol-based technologies does not mean we have to. We don't need to resort to blockchains in order to work together fairly. We are free to build distributed enterprises under the stewardship of traditional organizers utilizing centralized platforms. If anything, our mere ability to go elsewhere keeps the organizing body in check.

More important than the platform on which they are orchestrated, distributed enterprises give a capital-depleted population a way to invest and participate in a marketplace that all but ignores contributions other than cash. The work and resources required for production were already distributed among the people; they simply weren't recognized or accounted for as investment. The only part of the equation that could be effectively monopolized was the capital—which is why it was favored by the centralized power structures of industry.

As digital labor activist Trebor Scholz explains it, "They say that big money talks, but I say that platform cooperativism can invigorate genuine sharing, and that it does not have to reject the market. Platform cooperativism can serve as a remedy for the corrosive effects of capitalism; it can be a reminder that work can be dignified rather than diminishing for the human experience."[98]

That's how a digital economy *should* work. Instead of removing humans further from the equation of commerce, distributed digital technologies can reinvest living human beings into the fabric of a more sustainable and prosperous economic landscape.

There is no exit.

Chapter Five

DISTRIBUTED

DIGITAL DISTRIBUTISM

None of us can wave a magic wand and transform the economy from an extractive endgame into a prosperous commons. Most of us are struggling to make ends meet and, at best, trying to minimize the harm we cause in doing so. Oddly enough, many of us use what spare time and money we have earned *un*doing the impact of our jobs or companies. And that gets harder every day.

However, I do believe that once we've developed a sense of the direction in which we want the economy to go, we can reposition our careers and our businesses to become less parts of the problem than participants in the solution. There's a lot of hope here, if for no reason other than the fact that the best choices for your own and your business's future sustainability are the very same choices that make our collective economy more resilient and responsive to long-term human interests. Unlike industrialism, genuinely sustainable economics is not an either-or, zero-sum game.

As I hope I've shown, digital commerce can be a whole lot more than taking traditional corporate capitalism to the next level. Actually—or at least potentially—it's retrieving something much older and, to my mind, more positive for people and businesses alike. While the unimaginative big businesses of today see in the Internet a new way to automate labor, devalue human contributions, securitize wealth, build platform monopolies, and stage spectacular exits, stakeholders in tomorrow's economy should be able to see an opportunity to participate in self-sustaining, highly reciprocal, peer-to-peer, worker-owned, and community-defined marketplaces.

Instead of simply digitizing industrial extraction in the name of growing more capital, our new media technologies can *distribute* value creation in the name of a sustainable economy. It's time to move toward the fourth and final column in our chart.

	ARTISANAL 1000–1300	**INDUSTRIAL 1300–1990**	**DIGITAL INDUSTRIALISM 1990–2015**	**DIGITAL DISTRIBUTISM 2015–**
Direction	•	↗	⤴	⥀
Purpose	Subsistence	Growth	Exponential growth	Sustainable prosperity
Company	Family business	Chartered monopoly/ corporation	Platform monopoly (Amazon, Uber)	Platform cooperative (Mondragon, La'Zooz)
Currency	Market money (support trade)	Central currency (support banks)	Derivative instruments (leverage debt)	Bitcoin and P2P (promote circulation)
Investment	Direct investment	Stock markets	Algorithms	Crowdfunding
Production	Handmade (manuscript)	Mass-produced (printed book)	Replicable (file)	Collaborative (wiki)
Marketing	Human face	Brand icon	Big data (prediction)	Utility, legacy (product attributes, company ethics)
Communications	Personal contact	Mass media	Apps	Networks
Land & resources	Church commons	Colonization	Privatization	Public commons
Wages	Paid for value (craftsperson)	Paid for time (employee)	Not paid/underpaid (independent contractor)	Value exchanged (community member)
Scale	Local	National	Global	Strategically bounded
Optimized for	Creation of value	Extraction of value	Destruction of value	Exchange of value

As we saw, the artisanal economy was replaced by an industrial economy specifically geared to decrease the value of human contributions. Then we saw how each innovation and priority of the industrial era was amplified and intensified to new extremes by digital technology, leading to our current struggle with runaway and extractive digital industrialism.

But in the fourth column—what we'll call digital distributism—those mechanisms are reevaluated and reprogrammed to serve people and their businesses. Throughout this book, I've been offering some fledgling proposals for what those new approaches might look like. Digital industrialism sought to extract value from the system using new, digital means; digital distributism seeks to use those same technologies to distribute new capabilities to small businesses and real communities. Digital industrialism accepts growth as a condition of nature; digital distributism strives toward a dynamic steady state.

Where digital industrialism pushes corporations even further away from value creation, a more distributed approach to digital business embraces and enriches broader constituencies of stakeholders. Where an industrial approach to networking yields the platform monopolies of Uber and Amazon, a distributed one yields worker-owned cooperatives at a level of complexity and security unimaginable before digital technology. Where the digital industrialist's financial strategy is to extract money through increasingly abstracted derivatives, a more distributist vision would promote the circulation of money through low-friction, peer-driven currencies. Where digital industrialism seeks to use technology to expand markets forever, digital distributism seeks to recycle the same money again and again by investing and spending it in the bounded communities of the real world.

Where digital industrialism asks the economy to grow infinitely for its own sake, digital distributism aspires to sustainable prosperity. Such a steady state contradicts the growth-based economics of today's digital economy, not least because—unlike infinite growth—the goal of distributed wealth is actually *attainable*. We already enjoy more than enough surplus abundance to do this. We just have to be willing to accept widespread, stable prosperity as the purpose of the economy and then program for that.

To those who see the rules of a scarcity-based, industrial economy as fixed laws of nature, such an approach is heresy. This explains the disproportionate critique aimed at Pope Francis in the summer of 2015 for writing an encyclical (an official letter concerning Church doctrine) that made many of the same points I'm arguing here. On the surface, the pope's letter is encouraging the world to take global warming more seriously. But while he frames his argument as a plea to save the planet from human-caused climate change, he blames our economic system for bringing us to the brink of environmental collapse, and the alienating effects of digital technology for our inability to recognize this predicament. In brief, he argues that our belief in infinite growth is incompatible with a planet that has limited resources.

> It is not enough to balance, in the medium term, the protection of nature with financial gain, or the preservation of the environment with progress. . . . A technological and economic development which does not leave in its wake a better world and an integrally higher quality of life cannot be considered progress. Frequently, in fact, people's quality of life actually diminishes—by the deterioration of the environment, the low quality of food or the depletion of resources—in the midst of economic growth.[1]

Politicians balked—mostly because they had already been cynically using faith in God as a way of disproving global warming and freeing up the engines of capitalism to keep on churning no matter the apparent environmental impact. After all, how could human beings totally destroy something that God had created? By elevating human activity to the point where it could be blamed for global catastrophe, the pope had pulled the rug out from under the religious argument against climate change. "I think religion ought to be about making us better as people, less about things [that] end up getting into the political realm," responded climate change denier Jeb Bush. "I don't think we should politicize our faith."[2]

Even more infuriating to his critics than his warnings about climate change, I suspect, was the pope's critique of the industrial economy driving it. It sounded like the progressive, back-to-nature rhetoric of a co-op farmer or seed commons member:

> It is imperative to promote an economy which favours productive diversity and business creativity. For example, there is a great variety of small-scale food production systems which feed the greater part of the world's peoples, using a modest amount of land and producing less waste, be it in small agricultural parcels, in orchards and gardens, hunting and wild harvesting or local fishing. Economies of scale, especially in the agricultural sector, end up forcing smallholders to sell their land or to abandon their traditional crops. Their attempts to move to other, more diversified, means of production prove fruitless because of the difficulty of linkage with regional and global markets, or because the infrastructure for sales and transport is geared to larger businesses. Civil authorities have the right and duty to adopt clear and firm measures in support of small producers and differentiated production.[3]

On and on he went, explaining the ways that monopoly control of capital and resources not only leads to their overextraction but also prevents a majority of people from participating in value creation. If the pope's more distributed approach to land, labor, and capital seems informed by a larger economic logic, that's because it is. He didn't just pull this out of his mitre.

Although catalyzed by the excesses of our newly digital environment, the original ideals of distributism were first articulated, fittingly enough, in the encyclical letters of Pope Leo XIII[4] (1891) and Pius X[5] (1931), who were weighing in on the dangers of both Gilded Age capitalism and the rising tide of Marxism. Neither system is acceptable by itself, the popes explained, because while private property is indeed a human right (as capitalism argues), the resulting gross inequality is also immoral (as communism argues). They're both right. Instead of probing for a middle ground, however, the popes offered a radical alternative: to retrieve Catholicism's pre-Renaissance values of collective ownership and human agency.

Not surprisingly, the Vatican sees King Henry VIII's rejection of Catholicism and—more to the point—his "enclosure" of the commons as the moment when everything started to go in the wrong direction.* Hearkening back to an era before central currencies and chartered monopolies, the Church argued that the *factors of production* should once again be spread as widely as possible. Farmers should have access to land, craftsmen to tools, and, by inference today, digital creatives to the net. That's distributism, plain and simple: a sort of net neutrality applied to the whole world.

The popes saw that one individual's winnings were less important, in the collective long run, than everyone else's ability to make a living. While successful capitalists should be allowed to keep what they've earned, as winners they should not be able to monopolize the factors of production for themselves at the expense of their workers and peers. Or in today's parlance, while the founders of Amazon and Uber should be allowed to keep the money they make, they shouldn't be able to develop platform monopolies that disconnect workers from the resources they need to do their jobs or from earning an ownership stake in the platform itself. The ability to create and exchange value must remain distributed and available—a free market.

Early twentieth-century English writers Hilaire Belloc and G. K. Chesterton—and, later, a young Marshall McLuhan—saw in distributism a definitive answer to the failures of both capitalism and state socialism.[6, 7, 8] They looked to that same brief moment in the late Middle Ages we've

* And of course, widespread corruption within the Vatican itself played no small role in the collapse of the value system it was supposed to be promoting.

been exploring, when the market was in ascendance and former peasants were making and trading things, as the best example of the ideal economic system. Wealth was relatively widely dispersed, and people had a great deal of control over their livelihoods. They had access to the commons, to a low-cost marketplace, and to their own currencies and credit systems. Craftspeople belonged to trade guilds that both bounded their investment of labor and allowed for the advancement of skills to successive generations. The former peasants of this period became so collectively wealthy that they used their surplus profits to build cathedrals and municipal projects as investments in the future.

The centralization of power by the aristocracy and the great Renaissance that followed, according to all three popes, were less a pinnacle of human achievement than an undeserved celebration of dehumanizing technologies, economic injustice, colonial slavery, and an increasingly mechanized approach to life. In distributism, they saw a way to bring back what had been forcibly left behind by the industrial age and the rise of Protestant values that were, not coincidentally, much more directed toward personal achievement, individual wealth, and progress. But the manufacturing-based, highly industrial economy in which Leo XIII and Pius X lived just wasn't capable of supporting a postgrowth economic scheme geared toward equal opportunity and human prosperity. Besides, where was the real problem? Developing nations hadn't yet learned to effectively resist exploitation by wealthy ones, and the planet's environmental capabilities had yet to be tested.

If our current economic woes are not in themselves enough to motivate new approaches, digital technology with its distributed architecture may just encourage the sensibility and offer the infrastructure that a distributed economy requires. Computer chips and networks work by allocating tasks and sharing data. That's part of why it's so hard to keep our networks secure; they were built to share. That's also what makes them so good at promoting exchange: in a digital environment, everything gets distributed.

Distributism may be the best starting template for how to configure a digital economy. Unlike the expensive, centralized printing presses and warships of the industrial economy, digital technologies are intrinsically distributed. Distributed doesn't simply mean decentralized; it's not the principle through which alternative power centers emerge on the periphery of a system. Rather, when power is distributed, it is available throughout the network. It is everywhere at once. The same is true for capital in a distributed system, as well as value, energy, resources, companies, and people. Everything becomes more available to anyone.

New companies and those with established legacies alike can find in distributism a way to adapt to the highly collaborative yet limited growth landscape before us. It offers a strategy for individuals looking to make the transition from losing jobs to creating value in a peer-to-peer marketplace. And it suggests what governments can do to facilitate instead of hinder this transition to a postindustrial prosperity.

Distributism should not be confused with leftism. It's calling not for the *re*distribution of earnings or capital through taxes or state action after the fact but for the widest possible distribution of the means of production as *pre*conditions for a healthy marketplace. Workers ought to own the tools they use, and their contributions to an enterprise should earn them an ownership stake in the business itself. Distributism also discourages the externalization of costs to other parties or government, the privatization of currency, the treatment of economics as an impartial physical science, and the way big business and big government drain the market of liquidity.

Like the Internet itself, which spreads its packets through the path of least resistance without regard to hierarchy, distributism is to be guided by an economic principle the popes called *subsidiarity*.[9] Power is to be granted to the maximum number of the smallest possible nodes—guilds, communities, cottage businesses, and the family. This, in stark contrast not only to the self-interested libertarian individual but also to the seething mass of Communism's "the people." According to the principle of subsidiarity, no business should be bigger than it needs to be to serve its purpose—whether that's feeding pizza to the town or making roads for the state. Growth for growth's sake is discouraged. The ideal business, according to the popes, is the family business because of its limited size, its focus on long-term sustainability, and the likelihood that people will treat one another as something more dignified than replaceable employees. (And, as we've seen from the data, such businesses are also more resilient and long-lived.)

No, we don't need to convert to Catholicism or even approve of Vatican doctrine in order to appreciate the popes' vision of a more distributed economy and to see how it can contribute to our own. Besides, it's less their religious faith informing their recommendations than their memory of the wheels of commerce that preceded the engines of the industrial age. They are in many ways medievalists, after all, who can help us retrieve lost economic sensibilities the same way the Amish can remind us how to implement technology in a more considered fashion, or an aboriginal farmer can teach us how preindustrial crop rotation practices preserve soil nutrients. They remember.

A form of networked distributism may just be our last best hope for peace in the digital economy today. The conscious application of more distributist principles into the digital economic program could yield an entirely more prosperous and sustainable operating system. Instead of simply amplifying the most dehumanizing and extractive qualities of industrialism, it pushes ahead to something different—while also retrieving the truly free-market principles long obsolesced by corporatism.

RENAISSANCE NOW?

The digital industrialists have it wrong. There will be no second machine age. Like any truly new medium, digital technology will amplify and retrieve different values than the technologies that came before it did. Individuals and businesses will succeed in different ways and by different means than they have been.

This is good news: what was repressed by the industrial corporation can be retrieved and renewed by the distributed enterprise—while the excesses of the industrial era can themselves be repressed, or at least reduced, in a digital one. If anything, digital distributism retrieves some of the mechanisms and values of the artisanal era—that first column in the chart, back when everyone was growing, making, and trading stuff.

	ARTISANAL 1000–1300	DIGITAL DISTRIBUTISM 2015–
Direction	•	⥀
Purpose	Subsistence	Sustainable prosperity
Company	Family business	Platform cooperative (Mondragon, La'Zooz)
Currency	Market money (support trade)	Bitcoin and P2P (promote circulation)
Investment	Direct investment	Crowdfunding
Production	Handmade (manuscript)	Collaborative (wiki)
Marketing	Human face	Utility, legacy (product attributes, company ethics)
Communications	Personal contact	Networks
Land & resources	Church commons	Public commons
Wages	Paid for value (craftsperson)	Value exchanged (community member)
Scale	Local	Strategically bounded
Optimized for	Creation of value	Exchange of value

The platform cooperative recalls the values of the family business. Local currencies support the velocity of transactions, much like the market money of the bazaar. Crowdfunding lets people pay in advance for a business they want to see exist—in the fashion of an old village providing a barn and meals to a needed blacksmith until he could get his shop up and running. The handmade bias of the craft era becomes the hands-on bias of collaboration and wiki. The geographically local emphasis of the artisanal era is echoed in the consciously bounded investing and business practices of our era. Subsistence farming of the preindustrial era finds new expression in the sustainability agriculture agenda of today.

This leap forward by hearkening back is characteristic of any great cultural or economic shift. That's why at the vanguard of digital culture we see the retrieval of lost medieval values and practices, from the Burning Man festival to peer-to-peer currencies, and paganism to steampunk handcrafts. It's not moving backward. It doesn't constitute a regression so much as a recursion—a rediscovery of something old but in an entirely contemporary context. As Pope Francis put it, preempting this very accusation, "Nobody is suggesting a return to the Stone Age, but we do need to slow down and look at reality in a different way, to appropriate the positive and sustainable progress which has been made, but also to recover the values and the great goals swept away by our unrestrained delusions of grandeur."[10]

Keep the progress, but recover the lost values. Technically, then, he's talking about *renaissance:* the rebirth of old ideas in a new framework. It's a tall order, but given the profound reversals required to forge a new economic reality, a full-on renaissance may just be the goal here.

My late friend the philosopher Terence McKenna used to joke that if a person who knew nothing about human reproduction came upon a woman giving birth, he would surely think something terrible was going on: there's a giant tumor in her abdomen, there's screaming, there's even blood. He would conclude that the woman is dying. Yet we know the opposite is true. Though climactic, childbirth is not death but the emergence of new life.

Is it too hopeful to view the self-destructive bifurcation of our economy as, at least potentially, such a process? Might the extreme divisions of wealth we're enduring be less a permanent state than the sort of mitosis a cell undergoes just before it reproduces? Could a new economic landscape be emerging—a recovery of preindustrial mechanisms but enabled by digital platforms? Could this crisis be less our economy's death than its rebirth in a new form?

Let's put it to the test. If it's a genuine renaissance we're undergoing, we should be seeing evidence of widespread rebirth of lost values across many different sectors of society at once. For example, we remember the Renaissance for civilization-changing innovations such as perspective painting, circumnavigation of the globe, rational science, and the printing press. The Renaissance got its name because its many innovations "rebirthed" the long-obsolesced values of ancient Greece and Rome: the centrality of power and expansion of empire. If we're simply continuing the economics of the past, then we should expect to see those same values amplified in today's innovations. If we're enjoying a new renaissance, however, we should expect to see something else being reborn.

So what are today's parallels to the tremendous leaps made in the Renaissance? In the original Renaissance, the new technique of perspective painting gave artists the ability to represent three dimensions on a flat, two-dimensional surface. It was an amazing geometric trick that stunned the art world. But such pictures also conveyed a set of values. Perspective paintings required the viewer to stand at a particular angle, stressing the importance of a *single, correct perspective.*

Is there a digital era equivalent to perspective painting, which dazzles people with the illusion of increased dimensions? A hologram can represent a 3-D picture of a face or a bird, or even a moving image of, say, the face winking its eye or the bird flapping its wings. But in order to perceive the depth or motion of the image, observers must move across it. Instead of retrieving the Renaissance's emphasis on the single, objective, observing individual, a hologram recovers holism and relativity. The monopoly of a single perspective is distributed throughout the image. In fact, every tiny piece of a holographic plate—like a fractal, another digital corollary to perspective—contains information about the whole image. In contrast to the singular royal perspective retrieved by the Renaissance, we recover the more communal, even tribal value of clarity through collaboration.

Renaissance ships circumnavigated the globe, mapping the channels for colonial conquest. In our age, we have orbited the planet and photographed the tiny blue sphere from space. We have surrounded the globe with satellites and connected it with telecommunications. Where Renaissance man saw new territory to conquer and exploit, we see the finite reality of a biosphere to steward and protect. Spaceship Earth, the blue marble. We reclaim not the values of colonial conquistadors but those of a bounded community.

The literature and philosophy of the Renaissance retrieved the emphasis on an individual hero. This was the era of supreme individuals, such as Faustus, who sought to transcend mortality, Leonardo da Vinci's Godlike "Renaissance Man." They restored and upscaled the ancient Greek notion of a heroic "self." This emphasis on the individual, in turn, supported the values of self-interest and competitive economics. Our renaissance, on the other hand, is characterized by innovations such as computer gaming and social media. Storytelling today is less about individual authorship or heroes and more the collective fun of fantasy role-playing games and fanfiction sites. We don't read vicariously about a single character enduring his hero's journey; we join a massively multiplayer online role-playing game and make our own choices about how it will progress—in the virtual company of thousands of others. In the process, we train ourselves to make collective decisions about a distributed fantasy. Instead of retrieving the values of the court, we retrieve those of the folk.

Examples abound. Where the Renaissance initiated a system of arts patronage from above, funded by the aristocracy, our era promotes art from everywhere, funded by crowds. People's creative capacity has been unleashed irrespective of the taste of any elite. While the cause-and-effect logic of the Renaissance fostered a science of dissection, categories, and repeatability, the advent of genomics and computers promotes a new science of creation, design, and novelty. Finally, where the original Renaissance brought us the printing press, books, and the individual reader to own and interpret them, our era brings the Internet and—selfies notwithstanding—a newfound respect for networked intelligence and connectivity.

If we are, indeed, in a new renaissance—and I sincerely hope we are—then it only makes sense that we would reclaim some of the values that were repressed the last time we underwent such a shift. And so we're recovering lost cultural values from women's equality and integrative medicine to worker ownership and local currency. The Renaissance perpetuated the decline of handcrafted goods in favor of those manufactured by the machine and promoted by early corporate branding. Now we're back to limited-run artisanal beers and one-of-a-kind items. Our digital renaissance quite literally retrieves the *digits*—human fingers—as essential to the creation of value in a networked age.

Even if renaissance is a long shot, we may have no other option: like an overdue fetus becoming toxic to its mother, a new economy must be born or our very survival will be threatened. Economically driven climate change should be a big enough cue that it's time to go into labor; the polar shelf collapsing may as well be our water breaking. Indeed, history and humanity are both on the side of a new economy characterized less by industrial extraction than by digital distribution. And the planet appears to be demanding it, or else.

But how do we push toward such a wonderful outcome?

The many examples of more distributive business practices I've proposed throughout the book are, sadly, more exceptions than the rule. How are we regular workers and business owners, mortgage holders and real-estate brokers, supposed to usher in a better tomorrow or even just cope in the meantime? What can serve as a compass or North Star for navigating in this new terrain?

Let's use what we now know to develop one. Recall the tetrad we developed for the industrial corporation: it amplifies extraction, it obsolesces the peer-to-peer marketplace, it retrieves empire, and, when pushed to the extreme, it flips into personhood. What might the tetrad for a genuinely digital, distributist business look like? One that could live up to the demands of a renaissance?

It would amplify value creation from everywhere as much as from the center—distributed creativity. It would obsolesce centralized monopolies, working to break them up and share the means of production with customers. It would retrieve the values of the medieval marketplace, recovering inexpensive means of exchange between peers. Pushed to the extreme, well, a digitally distributed company would probably seek some sort of collective or spiritual awareness—another retrieval of a more familial or even tribal sensibility. (Where the traditional corporatists at Google seek to build an artificial intelligence and bring on the singularity, the digital-distributist equivalent would be to achieve a networked collective consciousness—but that's another book.)

Sounds pretty grandiose. But we can make our own personal and business decisions along those same lines. You don't have to be in charge of the company to argue for one set of priorities over another and to use the long-term health of the business and its market as your justification. An Internet platform can always lean away from monopoly and instead give its users the opportunity to create value for themselves. If you're working for the company behind it, you are entitled to make that argument, too. An airline can let fliers trade miles with one another. A restaurant can forgo the bank loan and crowdfund an expansion, just as a startup can choose not to scale but to grow as needed. A business can share ownership with its workers and let them participate in the expression of its mission. Independent contractors—competitors, even—can forge networked guilds and establish best practices and minimum wages. As investors we can begin to consider supporting companies that make our world a better place, and recognizing those returns as real. As consumers we can begin to value and reward the human labor invested in the things we buy.

In short, we use our digital sensibilities and technologies to retrieve and implement what was obsolesced by industrialism. *Digital* and *distributed* may even come to mean the same thing.

For now, however, we must soldier through the transition from one digital economy to another. Many people are retrofitting today's technologies to milk another century or two out of the last Renaissance while others are innovating optimistically toward the next one. This will have to do. No one can be faulted for maintaining a hybrid approach to an economy in such flux. Companies can't responsibly abandon their shareholders any more than individuals can abandon their IRAs. Compromise is not failure; it's incremental improvement in an imperfect situation.

I hope the principles I've laid out can serve as guidelines for where our activities and plans may fall on the spectrum and help us become more aware of which economy or economies we're operating in at any given moment. If we are measuring our health in terms of growth, then we are on the wrong track. If we are depending more on our competence, getting closer to value creation, allowing others to participate, investing in bounded communities where we actually live, and operating businesses we want to sustain instead of sell, then chances are we are moving in the right direction: grounded, collaborative, person-to-person exchange and support.

It's an economy we want to own.

Acknowledgments

This book is a product of the interactions I have had with hundreds of people over the past twenty years. I am grateful to every person who asked a question at a talk, e-mailed me about your situation, called in to a radio show, raised your hand in class, commented on an article, or tweeted me a link. Don't stop. I am: http://rushkoff.com, douglas@rushkoff.com, and @rushkoff on Twitter.

For implanting the dream of how a digital society and economy might function, I thank Internet cultural pioneers including Howard Rheingold, Mark Pesce, David Pescovitz, Mark Frauenfelder, Xeni Jardin, Cory Doctorow, John Barlow, Jaron Lanier, RU Sirius, Andrew Mayer, Richard Metzger, Evan Williams, everyone on the Well, Richard Stallman, George P'or, Neal Gorenflo, Marina Gorbis, and Michel Bauwens.

For leading digital enterprises in ways worth writing about, thanks to Scott Heiferman, Ben Knight, Zach Sims, Slava Rubin, the Robin Hood Cooperative, Enspiral, and Jimmy Wales. For sharing with me some of the perils of growth-based business and being open to discuss alternative possibilities, I thank Frank Cooper, Gerry Laybourne, Sara Levinson, Bonin Bough, Jon Kinderlerer, William Lohse, Ken Miller, and Judson Green.

Thanks to my publisher, Adrian Zackheim, for recognizing the single most important and counterintuitive assertion I'm making here, and to my editor, Niki Papadopoulos, for making sure it comes through loud and clear. Thanks to John Brockman for challenging me to tackle subjects that matter, and Katinka Matson for helping me craft my ideas into books and then finding them the right homes. Thanks to Jane Cavolina for bringing the full power of the English language to the service of these ideas, and Leah Trouwborst for turning the perfunctory into the delightful.

Thanks to the students who took my Digital Economics lab at ITP, for researching and workshopping many of these ideas, as well as innovating so many of your own. Thanks especially to Venessa Miemis, Adam Quinn, and Jon Wasserman for growing from brilliant students into inspiring colleagues. Thanks to Shareable, FastCoExist, Techonomy, the P2P Foundation, and the New Economy Coalition for sharing so many ideas, and to the funders and attendees of the Contact Summit I convened in 2011 for prototyping technological development outside the venture capital bubble. Thanks to everyone at Occupy for modeling alternative approaches to activism and at Burning Man for experimenting with new approaches to value exchange.

For consistently challenging my own assumptions about economics, I thank Bernard Lietaer, Brian Lehrer, Trebor Scholz, Amanda Palmer, and Micah Sifry. For consistently challenging my own assumptions about everything, I thank Seth Godin, Mark Stahlman, Amber Case, Mark Filippi, Joe Rogan, Marc Maron, and Helen Churko. Thanks to the Hermenautic Circle for your fellowship, and to Civic Hall for creating a physical space for this sort of discussion to occur as well as building a community willing to have it.

Thanks to Richard Maxwell, Mara Einstein, and the faculty of the Queens College Media Studies department for inviting me to help them build a graduate program for the study and practice of new solutions to the challenges of our political economy. Thanks to all the scholars and activists who have come to work with us already, and to those of you who may be encouraged to join after reading this book. You are invited: http://queenscollege.media.

Thanks to every community, company, and conference that invited me to speak or will in the future. It's where these ideas are born. And thanks to David Lavin and Charles Yao for making sure those opportunities happen. Thanks also to the independent bookstores, whose time has come again.

Thanks to Brian Hughes for assistance beyond reason.

Thanks most of all to my wife, Barbara, and daughter, Mamie, for keeping life fun.

Appendix

	ARTISANAL 1000–1300	INDUSTRIAL 1300–1990	DIGITAL INDUSTRIALISM 1990–2015	DIGITAL DISTRIBUTISM 2015–
Direction	•	↗	⤴	↻
Purpose	Subsistence	Growth	Exponential growth	Sustainable prosperity
Company	Family business	Chartered monopoly/ corporation	Platform monopoly (Amazon, Uber)	Platform cooperative (Mondragon, La'Zooz)
Currency	Market money (support trade)	Central currency (support banks)	Derivative instruments (leverage debt)	Bitcoin and P2P (promote circulation)
Investment	Direct investment	Stock markets	Algorithms	Crowdfunding
Production	Handmade (manuscript)	Mass-produced (printed book)	Replicable (file)	Collaborative (wiki)
Marketing	Human face	Brand icon	Big data (prediction)	Utility, legacy (product attributes, company ethics)
Communications	Personal contact	Mass media	Apps	Networks
Land & resources	Church commons	Colonization	Privatization	Public commons
Wages	Paid for value (craftsperson)	Paid for time (employee)	Not paid/underpaid (independent contractor)	Value exchanged (community member)
Scale	Local	National	Global	Strategically bounded
Optimized for	Creation of value	Extraction of value	Destruction of value	Exchange of value

Notes

Introduction: What's Wrong with This Picture?

1. Andrew Gumbel, "San Francisco's Guerrilla Protest at Google Buses Swells into Revolt," *Guardian*, January 25, 2014.
2. Vivian Giang, "A New Report Ranks America's Biggest Companies Based on How Quickly Employees Jump Ship," businessinsider.com, July 25, 2013.
3. Peter Schwartz and Peter Leyden, "The Long Boom: A History of the Future, 1980–2020," *Wired*, July 1997.
4. Telis Demos, Chris Dieterich, and Yoree Koh, "Twitter Shares Take Wing with Smooth Trading Debut," *Wall Street Journal*, November 6, 2013.
5. John Hagel et al., Foreword, "The Shift Index 2013: The 2013 Shift Index Series," Deloitte, 2013.

Chapter One: Removing Humans from the Equation

1. Eric S. Raymond, *The Cathedral and the Bazaar* (Sebastopol, Calif.: O'Reilly Media, 1999).
2. Women of the late Middle Ages in Europe were taller than at any other period until the 1970s. Bernard Lietaer and Stephen Belgin, *New Money for a New World* (Boulder, Colo.: Qiterra Press: 2011).
3. Douglas Rushkoff, *Life Inc.: How Corporations Conquered the World, and How We Can Take It Back* (New York: Random House, 2009), 8.
4. Erik Brynjolfsson and Andrew McAfee, *The Second Machine Age: Work, Progress, and Prosperity in a Time of Brilliant Technologies* (New York, London: W. W. Norton, 2014).

5. Michael Hauben and Ronda Hauben, "Netizen: On the History and Impact of Usenet and the Internet," *First Monday: Peer-Reviewed Journal on the Internet* 3, no. 7 (July 6, 1998).
6. "Organization: Organic," www.crunchbase.com/organization/organic#/entity.
7. Chris Anderson, *The Long Tail: Why the Future of Business Is Selling Less of More* (New York: Hyperion, 2006).
8. Anita Elberse, *Blockbusters: Hit-Making, Risk-Taking, and the Big Business of Entertainment* (New York: Henry Holt and Co., 2013).
9. Will Page and Eric Garland, "The Long Tail of P2P," *Economic Insight*, no. 14 (May 9, 2014).
10. Clay Shirky, "Power Laws, Weblogs and Inequality," shirky.com, February 8, 2003.
11. Elberse, *Blockbusters*.
12. Denise Lu, "Spotify vs. Bandcamp: Which Is More Band-Friendly?," mashable .com, November 19, 2013.
13. David Bollier, "The Tyranny of Choice: You Choose," economist.com, December 18, 2010.
14. Evelyn M. Rusli, "Facebook Buys Instagram for $1 Billion," nytimes.com, April 9, 2012.
15. Liz Gannes, "Instagram by the Numbers: 1 Billion Photos Uploaded," All Things D, April 3, 2012, allthingsd.com/20120403/instagram-by-the-numbers-1-billion -photos-uploaded/.
16. Hilary Heino, "Social Media Demographics—Instagram, Tumblr, and Pinterest," agileimpact.org, 2014.
17. Peter Cohan, "Yahoo's Tumblr Buy Fails Four Tests of a Successful Acquisition," forbes.com, May 20, 2013.
18. Chris Isidore, "Yahoo Buys Tumblr, Promises Not to 'Screw It Up,'" money.cnn .com, May 20, 2013.
19. Jordan Crook, "Snapchat Sees More Daily Photos Than Facebook," techcrunch .com, November 19, 2013.
20. Vindu Goel, "Facebook Tinkers with Users' Emotions in News Feed Experiment, Stirring Outcry," nytimes.com, June 29, 2014.
21. Astra Taylor, *The People's Platform: The Culture of Power in a Networked Age* (New York: Henry Holt and Co., 2014), 205.
22. "Generation Like," *Frontline*, PBS, February 18, 2014.
23. Bob Lefsetz, "Tom Petty," *The Lefsetz Letter*, July 22, 2014.
24. Wire staff, "AOL Acquires Huffington Post for $315 Million," money.cnn.com, February 7, 2011.
25. "Generation Like," *Frontline*.
26. Jon Pareles, "Jay-Z Is Watching and He Knows Your Friends," nytimes.com, July 4, 2013.

27. According to Galbithink.org ("Annual U.S. Advertising Spending Since 1919," September 14, 2008), total ad and marketing spend in the United States has remained at less than 3 percent of GDP through the print, radio, television, and Internet eras. In 1925, advertising and marketing as a share of GDP was approximately 2.9 percent. By 1998, it was actually down to about 2.4 percent.

    Market research, even counted separately (which likely counts these amounts—normally baked into ad budgets—twice), only accounts for about 0.12 percent more. Adding to this all additional promotions, research, demographics, direct marketing, and public relations doesn't even double the figures. For the best data, see Glen Wiggs, "AdSpend and GDP—2014 Update," Foundation for Advertising Research, June 19, 2014. Also, see U.S. Census and IRS data for advertising and marketing expenditures, which is well parsed and linked by Douglas Galbi at the Purple Motes Web site, purplemotes.net/2008/09/14/us-advertising-expenditure-data/, accessed January 2015.

    According to the marketing industry's own publication, *Advertising Age Marketing Fact Pack, 2015 Edition* (New York: Crain, 2014), the total global advertising spend for 2015 was to be an estimated $545 billion. With gross world product (GWP) at about $75 trillion, that represents 0.7 percent.
28. Dominic Rushe, "Nearly 25% of 'People' Viewing Online Video Ads Are Robots Used by Fraudsters," theguardian.com, December 9, 2014.
29. Tracy Boyer Clark, "Wall Street Journal's Digital Strategy Amidst the Digital Revolution," innovativeinteractivity.com, May 14, 2012.
30. Ryan Chittum, "The Upside of Yesterday's *New York Times* News," cjr.com, October 2, 2014.
31. Joshua Clover, "Amanda Palmer's Accidental Experiment with Real Communism," newyorker.com, October 2, 2012.
32. Natasha Singer, "Mapping, and Sharing, the Consumer Genome," nytimes.com, June 16, 2012.
33. Brian Womack, "Google Updates Flu Trends to Improve Accuracy," businessweek.com, November 1, 2014.
34. Jaron Lanier, *Who Owns the Future?* (New York: Simon and Schuster, 2013), 286.
35. Ibid., 227.
36. Ibid., 20.
37. Jason Clampet, "Airbnb in NYC: The Real Numbers Behind the Sharing Story," skift.com, February 13, 2014.
38. Ron Miller, "An Uber Valuation Comes with Uber Problems," techcrunch.com, December 16, 2014.
39. "Organization: Uber," www.crunchbase.com/organization/uber.
40. Moshe Z. Marvit, "How Crowdworkers Became the Ghosts in the Digital Machine," thenation.com, February 4, 2014.

41. Trebor Scholz, "Crowdmilking," collectivate.net, March 9, 2014.
42. Andrew Keen, *The Internet Is Not the Answer* (New York: Atlantic Monthly Press, 2015).
43. Vivek Wadhwa, "The End of Chinese Manufacturing and Rebirth of U.S. Industry," forbes.com, July 23, 2012.
44. Daniel Bell, *The Coming of Post-Industrial Society: A Venture in Social Forecasting* (New York: Basic Books, 1976).
45. David Rotman, "How Technology Is Destroying Jobs," technologyreview.com, June, 12, 2013.
46. Ibid.
47. Thomas Piketty, *Capital in the Twenty-First Century*, trans. Arthur Goldhammer (Cambridge, Mass.: Belknap Press, 2014).
48. Bernard Lietaer, *The Mystery of Money: Beyond Greed and Scarcity*, 148 [PDF].
49. Jeff Tyler, "Banks Demolish Foreclosed Homes, Raise Eyebrows," *Marketplace*, American Public Media, October 13, 2011. Transcript available at www.marketplace.org/topics/business/banks-demolish-foreclosed-homes-raise-eyebrows/.
50. Dr. Jacques Diouf, "Towards a Hunger-Free Century," Millennium Lecture, M. S. Swaminathan Foundation, Chennai (Madras), India, April 29, 1999, available at Food and Agriculture Organization of the United Nations, FAO.org.
51. Juliet B. Schor, *The Overworked American: The Unexpected Decline of Leisure* (New York: Basic Books, 1992).
52. Juliet Schor and Julia Slay, "Attitudes About Work Time and the Path to a Shorter Working Week," New Economics Institute, Strategies for a New Economy Conference, June 2012, vimeo.com/47179682.
53. Juliet B. Schor, *Plenitude: The New Economics of True Wealth* (New York: Penguin Press, 2010).
54. "Telecommuters with Flextime Stay Balanced up to 19 Hours Longer," accounting.smartpros.com, July 2010; Daniel Cook, "Rules of Productivity Presentation," lostgarden.com, September 28, 2008.
55. Jenny Brundin, "Utah Finds Surprising Benefits in 4-Day Workweek," npr.org, April 10, 2009.
56. John de Graaf, "Life Away from the Rat-Race: Why One Group of Workers Decided to Cut Their Own Hours and Pay," alternet.org, July 2, 2012.
57. Bryce Covert, "This Company Has a 4-Day Work Week, Pays Its Workers a Full Salary and Is Super Successful," thinkprogress.org, April 18, 2014.
58. Ilya Pozin, "Thursday Is the New Friday: Embracing the 4-Day Work Week," linkedin.com, May 6, 2014.
59. Reid Hoffman, *The Alliance: Managing Talent in the Networked Age* (Cambridge, Mass.: Harvard Business Review Press, 2014).
60. Jeremy Rifkin, *The Zero Marginal Cost Society* (Basingstoke, England: Palgrave Macmillan, 2014).

61. Mike Alberti and Kevin C. Brown, "Guaranteed Income's Moment in the Sun," remappingdebate.org, October 3, 2013.
62. Ibid.
63. Matt Bruenig, "This One Weird Trick Actually Cuts Child Poverty in Half," demos.org, July 21, 2014.
64. Dylan Matthews, "A Guaranteed Income for Every American Would Eliminate Poverty—And It Wouldn't Destroy the Economy," vox.com, July 23, 2014.
65. Anthony B. Atkinson, *Inequality: What Can Be Done?* (Cambridge, Mass.: Harvard University Press, 2015).

Chapter Two: The Growth Trap

1. Marshall McLuhan and Eric McLuhan, *Laws of Media: The New Science* (Toronto: University of Toronto Press, 1992).
2. For the best historical explanation of the negative effect of chartered commerce and corporatism on the marketplace, see Fernand Braudel, *Civilization and Capitalism, 15th–18th Century*, vol. 1: *The Structures of Everyday Life* (New York: Harper & Row, 1982).
3. Ann M. Carlos and Stephen Nicholas, "'Giants of an Earlier Capitalism': The Chartered Trading Companies as Modern Multinationals," *Business History Review* 62, no. 3 (1988): 398–419.
4. Thom Hartmann, *Unequal Protection: The Rise of Corporate Dominance and the Theft of Human Rights* (New York: Rodale Books, 2002).
5. Binyamin Appelbaum, "What the Hobby Lobby Ruling Means for America," nytimes.com, July 22, 2014.
6. Douglas Rushkoff, *Life Inc.: How Corporations Conquered the World, and How We Can Take It Back* (New York: Random House, 2009), 13.
7. *Store Wars: When Wal-Mart Comes to Town*, PBS, directed by Micha X. Peled (2001).
8. Abigail Goldman, "Wal-Mart not joking with smiley face lawsuit," *Los Angeles Times*, May 14, 2006.
9. "Walmart," www.lippincott.com/en/work/walmart.
10. "Is Walmart Good for America?," *Frontline*, PBS, November 16, 2004.
11. Rick Ungar, "Walmart Pays Workers Poorly and Sinks While Costco Pays Workers Well and Sails—Proof That You Get What You Pay For," forbes.com, April 17, 2013.
12. John Hagel et al., "Measuring the Forces of Long-term Change: The 2009 Shift Index," Deloitte, 2009.
13. John Hagel et al., Foreword, "The Shift Index 2013: The 2013 Shift Index Series," Deloitte, 2013.
14. John Hagel, John Seely Brown, and Duleesha Kulasooriya, *Shift Happens: How the World Is Changing, and What You Need to Do About It* (Houston, Tex.: Idea Bite Press, 2014).

15. Hagel et al., Foreword, "The Shift Index 2013."
16. "'Trying to Recapture the Magic': The Strategy Behind the Pharma M&A Rush," knowledge.wharton.upenn.edu, May 28, 2014.
17. Rushkoff, *Life Inc.*, 174.
18. Beth Ann Bovino et al., "How Increasing Income Inequality Is Dampening Economic Growth, and Possible Ways to Change the Tide," globalcreditreport.com, August 5, 2014.
19. Geoffrey Rogow, "Wealth Inequality Can Damage Economy, S&P's Bovino Says," blogs.wsj.com, August 5, 2014.
20. Joseph A. Schumpeter, *Capitalism, Socialism and Democracy*, 3rd ed. (New York: Harper Perennial, 2008).
21. "Worldwide Revenue of Major Toy Companies in 2012 (in Million U.S. Dollars)," statista.com, 2015.
22. *Media Squat*, WFMU, June 8, 2009.
23. Megan Rose Dickey, "We Talked to Uber Drivers—Here's How Much They Really Make," businessinsider.com, July 18, 2014.
24. Aaron Sankin, "Why New York Taxis Are Powerless Against Uber's Price War," dailydot.com, July 8, 2014.
25. Don Jergler, "Transportation Network Companies, Uber Liability Gap Worry Insurers," insurancejournal.com, February 10, 2014.
26. Tim Bradshaw, "Uber's Tactics Pay Off as It Goes Head to Head with US Rival," ft.com, September 11, 2014.
27. Fred Wilson, "Platform Monopolies," avc.com, July 13, 2014.
28. David Streitfeld, "Amazon, a Friendly Giant as Long as It's Fed," nytimes.com, July 12, 2014.
29. Venkatesh Rao, "Why Amazon Is the Best Strategic Player in Tech," forbes.com, December 14, 2011.
30. Elvis Picardo, "Apple? Google? Tesla? Which Will Be the First to Reach a $1 Trillion Market Cap?," investopedia.com, July 7, 2014.
31. Wilson, "Platform Monopolies."
32. Josh Constine, "Hail a Fellow Human, Not a Taxi with 'Sidecar'—The New P2P Uber," techcrunch.com, June 26, 2012.
33. Wilson, "Platform Monopolies."
34. Betsy Corcoran, "Blackboard's Jay Bhatt Strikes Up the Brass Band," edsurge.com, July 23, 2014.
35. Justin Pope, "E-Learning Firm Sparks Controversy with Software Patent," washingtonpost.com, October 15, 2006.
36. withknown.com.
37. Carlota Perez, *Technological Revolutions and Financial Capital* (Cheltenham, England: Edward Elgar Press, 2002).

38. Thomas Piketty, *Capital in the Twenty-First Century,* trans. Arthur Goldhammer (Cambridge, Mass.: Belknap Press, 2014).
39. Mario Preve, quoted in Ernst & Young and Family Business Network International, "Built to Last: Family Businesses Lead the Way to Sustainable Growth" (n.p.: Ernst & Young Global Limited, 2012), www.ey.com/Publication/vwLUAssets/EY-Built-to-last-family-businesses-lead-the-way-to-sustainable-growth/$FILE/EY-Built-to-last-family-businesses-lead-the-way-to-sustainable-growth.pdf.
40. Nicolas Kachaner, George Stalk, and Alain Bloch, "What You Can Learn from Family Business," *Harvard Business Review,* November 2012.
41. Yoko Kubota and Maki Shiraki, "After Two Bumper Years, Toyota Braces for Shift to Slower Growth," reuters.com, April 15, 2014.
42. Ibid.
43. pgconnectdevelop.com.
44. Larry Huston and Nabil Sakkab, "Connect and Develop: Inside Procter & Gamble's New Model for Innovation," *Harvard Business Review,* March 2006.
45. "Connect + Develop History," ConnectDevelop, YouTube, August 16, 2012.
46. Huston and Sakkab, "Connect and Develop."
47. "Procter & Gamble Re-ignites Growth—XBD & Open Innovation Make It Happen," federicibusiness.com, 2008.
48. "Febreze Embracing C+D to Become a Billion $ Brand," pgconnectdevelop.com, January 1, 2013.
49. Clark Gilbert, Matthew Eyring, and Richard N. Foster, "Two Routes to Resilience," *Harvard Business Review,* December 2012.
50. Field Maloney, "Is Whole Foods Wholesome?" slate.com, March 17, 2006.
51. Lynn Forester de Rothschild, "Capitalists for Inclusive Growth," project-syndicate.org, April 17, 2013.
52. Ibid.
53. Ibid.
54. Steven Pearlstein, "How the Cult of Shareholder Value Wrecked American Business," washingtonpost.com, September 9, 2013.
55. Oliver Staley and Hui-Yong Yu, "Hilton Sells Itself to Blackstone for $20 Billion," bloomberg.com, July 4, 2007.
56. Henry Sender, "How Blackstone Revived Hilton Brand," ft.com, August 19, 2013.
57. David Gelles, "A Surprise from Hilton: Big Profit for Blackstone," nytimes.com, December 12, 2013.
58. Nanette Byrnes and Peter Burrows, "Where Dell Went Wrong," businessweek.com, February 18, 2007.
59. Ashlee Vance, "Why Michael Dell Really Had to Take Dell Private," businessweek.com, February 5, 2013.

60. Mary Ellen Biery, "Why Michael Dell's Fight Makes Sense," forbes.com, August 11, 2013.
61. Ibid.
62. Connie Guglielmo, "Dell Officially Goes Private: Inside the Nastiest Tech Buyout Ever," forbes.com, October 30, 2013.
63. Lindsey Rupp, Carol Hymowitz, and David Carey, "Drexler Amasses $350 Million as J. Crew Struggles," businessweek.com, June 13, 2014.
64. Ibid.
65. Lindsey Rupp, "J. Crew Profits Fall as Company Considers Going Public Again," bloomberg.com, March 25, 2014.
66. Paula Kepos, *International Directory of Company Histories*, vol. 7 (Farmington Hills, Mich.: St. James Press, 1993), per "Amsted Industries Incorporated History" entry on fundinguniverse.com.
67. "The Employee Ownership 100: America's Largest Majority Employee-Owned Companies," National Center for Employee Ownership, nceo.org, June 2014.
68. Brian Solomon, "The Wal-Mart Slayer: How Publix's People-First Culture Is Winning the Grocer War," forbes.com, July 24, 2013.
69. Sabri Ben-Achour, "Groceries: A Low Margin Business, but Still Highly Desirable," marketplace.org, September 12, 2013.
70. Derek Ridgway, "Flexible Purpose Corporation vs. Benefit Corporation," hanson bridgett.com, September 4, 2012.
71. "Inc. 5000," inc.com, September 6, 2013.
72. Ariel Schwartz, "Inside Plum Organics, the First Benefit Corporation Owned by a Public Company," fastcoexist.com, January 22, 2014.
73. Marc Gunther, "Checking In with Plum Organics, the Only B Corp Inside a Publicly Traded Company," theguardian.com, August 6, 2014.
74. Ridgway, "Flexible Purpose Corporation vs. Benefit Corporation."
75. Kyle Westaway, "PROFIT + PURPOSE—Structuring Social Enterprise for Impact," slideshare.net, March 6, 2012.
76. Cameron Scott, "Tiny AI Startup Vicarious Says It's Solved CAPTCHA," singu larityhub.com, October 29, 2013.
77. vicarious.com/about.html.
78. Westaway, "PROFIT + PURPOSE."
79. Citizen Media Law Project, "Primer on Low-Profit Limited Liability Companies (L3Cs)," Berkman Center for Internet & Society at Harvard University, October 2010.
80. homeportneworleans.org.
81. battle-bro.com/.
82. Donnie Maclurcan and Jennifer Hinton, "Beyond Capitalism: Not-for-Profit Business Ethos Motivates Sustainable Behaviour," theguardian.com, October 1, 2014.

83. “Exemption Requirements—501(c)(3) Organizations,” irs.gov, January 8, 2015.
84. John Tozzi, “Turning Nonprofits into For-Profits,” businessweek.com, June 15, 2009.
85. “Mozilla Foundation Announces Creation of Mozilla Corporation,” mozillazine .org, August 3, 2005, per Wayback Machine at archive.org/web/.
86. “Articles of Incorporation of M. F. Technologies,” static.mozilla.com, July 14, 2003.
87. www.linkedin.com/company/mozilla-corporation, 2015.

Chapter Three: The Speed of Money

1. U.S. Department of the Treasury, “History of ‘In God We Trust,’” treasury.gov, March 8, 2011.
2. Christopher Simpson, *Science of Coercion: Communication Research & Psychological Warfare 1945–1960* (Oxford: Oxford University Press, 1996).
3. Luca Fantacci, “The Dual Currency System of Renaissance Europe,” *Financial History Review* 15, no. 1 (2008).
4. Carlo M. Cipolla, *Before the Industrial Revolution: European Society and Economy, 1000–1700,* 3rd ed. (New York: W. W. Norton, 1994).
5. Ibid.
6. Ibid.
7. Douglas Rushkoff, *Life Inc.: How Corporations Conquered the World, and How We Can Take It Back* (New York: Random House, 2009), 164.
8. Ibid., 8–10.
9. Ibid., 167–70.
10. Bernard A. Lietaer and Stephen M. Belgin, *Of Human Wealth: Beyond Greed & Scarcity* (Boulder, Colo.: Human Wealth Books and Talks, 2001), 111.
11. For a more detailed discussion of this basic principle, see Thomas H. Greco, *Understanding and Creating Alternatives to Legal Tender* (White River Junction, Vt.: Chelsea Green Publishing Co., 2001).
12. Rushkoff, *Life Inc.*, 170–71.
13. Michael Konczal, “Frenzied Financialization,” washingtonmonthly.com, November/December 2014.
14. Thomas Piketty, *Capital in the Twenty-First Century*, trans. Arthur Goldhammer (Cambridge, Mass.: Belknap Press, 2014).
15. Robert Slater, *Jack Welch and the GE Way* (New York: McGraw-Hill, 1998).
16. Ben Steverman, “Manipulate Me: The Booming Business in Behavioral Finance,” bloomberg.com, April 7, 2014.
17. Morgan House, “5 Alan Greenspan Quotes That Make You Wonder,” fool.com, October 15, 2008.
18. Michael Lewis, *The Big Short: Inside the Doomsday Machine* (New York, London: W. W. Norton, 2011).

19. Naomi Klein, *This Changes Everything: Capitalism vs. the Climate* (New York: Simon and Schuster, 2014).
20. John Stuart Mill, *Principles of Political Economy with Some of Their Applications to Social Philosophy* (London: Longmans, Green and Co., 1909), IV.6.2.
21. Ibid., IV.6.7.
22. David Dayen, "America's Ugly Economic Truth: Why Austerity Is Generating Another Slowdown," salon.com, October 21, 2014.
23. David Wessel, "Lousy Economic Growth Is a Choice, Not an Inevitability," brookings.edu, October 13, 2014.
24. Bernard Lietaer and Jacqui Dunne, *Rethinking Money: How New Currencies Turn Scarcity into Prosperity* (San Francisco: Berrett-Koehler Publishers, 2013).
25. Joanna Glasner, "PayPal's IPO Woes Continue," *Wired*, February 12, 2002.
26. In most of the world, that would be SWIFT.
27. Satoshi Nakamoto, "Bitcoin: A Peer-to-Peer Electronic Cash System," bitcoin.org, October 31, 2008.
28. Ibid.
29. Pedro Franco, *Understanding Bitcoin: Cryptography, Engineering and Economics* (New York: John Wiley & Sons, 2014).
30. Ibid.
31. Andreas M. Antonopoulos, *Mastering Bitcoin: Unlocking Digital Cryptocurrencies* (Sebastopol, Calif.: O'Reilly Media, 2014).
32. Franco, *Understanding Bitcoin.*
33. Antonopoulos, *Mastering Bitcoin.*
34. Rob Wile, "The Chinese Are in Love with Bitcoin and It's Driving the Digital Currency's Prices into the Stratosphere," businessinsider.com, October 29, 2013.
35. Rebecca Grant, "A Single Bitcoin Was Worth $10 a Year Ago—Today It's Worth $1,000," venturebeat.com, November 27, 2013.
36. Robert McMillan, "The Inside Story of Mt. Gox, Bitcoin's $460 Million Disaster," wired.com, March 3, 2014.
37. Ryan Lawler, "Bitcoin Miners Are Racking Up $150,000 a Day in Power Consumption Alone," techcrunch.com, April 13, 2013.
38. Mark Gimein, "Virtual Bitcoin Mining Is a Real-World Environmental Disaster," bloomberg.com, April 12, 2013.
39. Michael Carney, "Bitcoin Has a Dark Side: Its Carbon Footprint," pando.com, December 16, 2013.
40. Lawler, "Bitcoin Miners Are Racking Up $150,000 a Day."
41. Jon Evans, "Enter the Blockchain: How Bitcoin Can Turn the Cloud Inside Out," techcrunch.com, March 22, 2014.
42. Vitalik Buterin, "DAOs, DACs, DAs and More: An Incomplete Terminology Guide," blog.ethereum.org, May 6, 2014.

43. David Johnston, Sam Onat Yilmaz, Jeremy Kandah, Nikos Bentenitis, Farzad Hashemi, Ron Gross, Shawn Wilkinson, and Steven Mason, "The General Theory of Decentralized Applications, Dapps," github.com, June 9, 2014.
44. Nakamoto, "Bitcoin: A Peer-to-Peer Electronic Cash System."
45. National Patient Advocate Foundation, "Issue Brief: Medical Debt, Medical Bankruptcy and the Impact on Patients," npaf.org, September 2012.
46. Dan Mangan, "Medical Bills Are the Biggest Cause of US Bankruptcies: Study," cnbc.com, June 25, 2013.
47. National Patient Advocate Foundation, "Issue Brief: Medical Debt, Medical Bankruptcy and the Impact on Patients."
48. rollingjubilee.org.
49. Interview with Astra Taylor, cofounder of Strike Debt and the Rolling Jubilee, conducted by e-mail, July 24, 2015.
50. "A Look Back at the 2012 ABA Indie Impact Study Series," localismbythenumbers.com, June 11, 2014.
51. Justin Sacks, *The Money Trail: Measuring Your Impact on the Local Economy Using LM3* (London: New Economics Foundation & the Countryside Agency, December 2002).
52. "A Look Back at the 2012 ABA Indie Impact Study Series."
53. Bill McKibben, "A Day in the Life of a BerkShare," yesmagazine.org, October 18, 2010.
54. Katie Gilbert, "Why Local Currencies Could Be on the Rise in the U.S.—And Why It Matters," forbes.com, September 22, 2014.
55. "Group Hopes 'Detroit Dollar' Pays Off for Biz," crainsdetroit.com, March 16, 2014.
56. John Rogers, "Bristol Pound Is Just One Example of What Local Currencies Can Achieve," theguardian.com, June 17, 2013.
57. Lietaer and Dunne, *Rethinking Money*, 175–81.
58. Lietaer and Dunne, *Rethinking Money*.
59. Ibid.
60. Irving Fisher, *The Purchasing Power of Money* (New York: Macmillan, 1920).
61. Irving Fisher, *Stamp Scrip* (New York: Adelphi, 1933).
62. Loren Gatch, "Local Money in the United States During the Great Depression," *Essays in Economics & Business History* 26 (2008).
63. Ibid.
64. Lauren Frayer, "'Time Banks' Help Spaniards Weather Financial Crisis," npr.org, September 22, 2012.
65. See TimeBanks USA at timebanks.org for a new time dollars smartphone app, or p2pfoundation.net/Complementary_Currency_Software for a comprehensive list of complementary currency software.

66. Bernard Lietaer, *The Future of Money: Creating New Wealth, Work and a Wiser World* (London: Random House, 2001).
67. Mayumi Hayashi, "Japan's Fureai Kippu Time-Banking in Elderly Care: Origins, Development, Challenges, and Impact," *International Journal of Community Currency Research* 15 (2012).
68. Ariana Eunjung Cha, "In Spain, Financial Crisis Feeds Expansion of a Parallel, Euro-Free Economy," washingtonpost.com, August 27, 2012.
69. Lietaer and Dunne, *Rethinking Money*, 143.
70. Ibid., 142.
71. Ibid., 143.

Chapter Four: Investing Without Exiting

1. Helaine Olen, *Pound Foolish: Exposing the Dark Side of the Personal Finance Industry* (New York: Penguin/Portfolio, 2012).
2. OECD, *Protecting Pensions: Policy Analysis and Examples from OECD Countries* (Paris: OECD Publications, 2007), 268.
3. Steve Wilhelm, "Why Boeing's Fighting to Retire Pensions," bizjournal.com, January 11, 2013.
4. John W. Miller, "Steelmaker Presses for 36% Pay Cut," wsj.com, July 20, 2012.
5. James R. Hagerty and Alistair MacDonald, "As Unions Lose Their Grip, Indiana Lures Manufacturing Jobs," wsj.com, March 18, 2012.
6. Associated Press, "10 Years Later: What Happened to the Former Employees of Enron?," businessinsider.com, December 1, 2011.
7. Chris Gay, "The 401(k)'s 'Father' Wants to Hit Reset," money.usnews.com, September 20, 2012.
8. Olen, *Pound Foolish*, 81.
9. Ibid., 85.
10. Mitch Tuchman, "Pension Plans Beat 401(k) Savers Silly—Here's Why," forbes.com, June 4, 2013.
11. Olen, *Pound Foolish*, 89.
12. William E. Even and David A. Macpherson, "Why Did Male Pension Coverage Decline in the 1980s?," *Industrial and Labor Relations Review* 47, no. 3 (April 1993).
13. Olen, *Pound Foolish*, 82.
14. "Pensions Decline as 401(k) Plans Multiply," bankrate.com, July 24, 2014.
15. Olen, *Pound Foolish*, 85.
16. Gretchen Morgenson, "The Curtain Opens on 401(k) Fees," nytimes.com, June 2, 2012.
17. Tuchman, 2013.
18. Olen, *Pound Foolish*, 86.
19. Ibid., 98.

20. Michael Shuman, *Local Dollars, Local Sense* (White River Junction, Vt.: Chelsea Green Publishing Co.), 2012.
21. "Sell Your Stocks, MIT Sloan Professor Urges Small Investors Saving for Retirement," mitsloan.mit.edu, March 12, 2009.
22. Bob Wallace, "AT&T Service Helps Broker Shave Costs," *Network World* 7, no. 31 (July 30, 1990).
23. Martin LaMonica, "Bullish on the Net," *InfoWorld*, April 26, 1999: 34–35.
24. Riva D. Atlas, "Trading Slump Spurs Online Brokers' Merger Talk," nytimes.com, May 10, 2005.
25. Brad M. Barber and Terrance Odean, "The Internet and the Investor," *Journal of Economic Perspectives* 15, no. 1 (Winter 2001): 41–54.
26. Joe Light and Julie Steinberg, "Small Investors Jump Back into the Trading Game," wsj.com, February 21, 2014.
27. D. K. Peterson and G. F. Pitz, "Confidence, Uncertainty, and the Use of Information," *Journal of Experimental Psychology: Learning, Memory and Cognition* 14 (1988): 85–92, as cited in Barber and Odean, "The Internet and the Investor."
28. Barber and Odean, "The Internet and the Investor."
29. Ibid.
30. Ibid.
31. Richard Finger, "High Frequency Trading: Is It a Dark Force Against Ordinary Human Traders and Investors?," forbes.com, September 30, 2013.
32. Ibid.
33. Jerry Adler, "Raging Bulls: How Wall Street Got Addicted to Light-Speed Trading," wired.com, August 3, 2012.
34. Simone Foxman, "How the 'Navy SEALS' of Trading Are Taking on Wall Street's Predatory Robots," qz.com, March 31, 2014.
35. Alan Kohler, "$710 trillion: That's a Lot of Exposure to Derivatives," abc.net.au, June 11, 2014.
36. Central Intelligence Agency, *The World Factbook*, www.cia.gov/library/publications/the-world-factbook/geos/xx.html.
37. Kohler, "$710 trillion: That's a Lot of Exposure to Derivatives."
38. Peter Cohan, "Big Risk: $1.2 Quadrillion Derivatives Market Dwarfs World GDP," DailyFinance (AOL), June 9, 2010, www.dailyfinance.com/2010/06/09/risk-quadrillion-derivatives-market-gdp/.
39. Christopher Matthews, "Why the New York Stock Exchange Sold Out to an Upstart You've Never Heard Of," business.time.com, December 21, 2012.
40. Herman Daly, *Beyond Growth: The Economics of Sustainable Development* (Boston: Beacon Press, 1996), 37.
41. GarageGames (huge gaming platform), Sierra Entertainment (makers of *King's Quest*), Apple, and Dell, to name just a few.

42. "Brain Maturity Extends Well Beyond Teen Years," *Tell Me More*, NPR, October 10, 2011.
43. Kathleen De Vere, "Draw Something Surpasses 50 Million Downloads, May Have as Many as 24 Million Daily Active Users," adweek.com, April 4, 2012.
44. Julianne Pepitone, "Zynga IPO Values Company at $7 Billion," money.cnn.com, December 16, 2011.
45. Josh Constine, "Zynga Shares Go on Wild Ride During Facebook IPO—Big Fall, Then Recovery," techcrunch.com, May 18, 2012.
46. Sarah McBride and Leana B. Baker, "Zynga Buys OMGPOP Games Company for $200 Million: Source," reuters.com, March 21, 2012.
47. Paul Tassi, "Draw Something Loses 5M Users a Month After Zynga Purchase," forbes.com, May 4, 2012.
48. Sam Biddle, "OMGPOP Is Dead," valleywag.gawker.com, June 4, 2013.
49. Ari Levy, "Google Shares Took Off, but the Auction Didn't," cnbc.com, August 19, 2014.
50. "Google IPO Priced at $85 a Share," edition.cnn.com, August 19, 2004.
51. Kevin J. Delaney and Robin Sidel, "How Miscalculations and Hubris Hobbled Celebrated Google IPO," wsj.com, August 19, 2004.
52. Interview with Scott Heiferman, conducted by e-mail, September 2014.
53. Sarah Lacy, "Pando in 2014: Looking Back on an Exhausting, Transformational Year," pando.com, December 25, 2014.
54. Max Chafkin, "True to Its Roots: Why Kickstarter Won't Sell," fastcompany.com, March 18, 2013.
55. "Kickstarter Is a Benefit Corporation," kickstarter.com, September 21, 2015.
56. J. D. Alois, "Neil Young's Pono Music Is Now Equity Crowdfunding Following $6.2 Million Kickstarter Hit," crowdfundinsider.com, August 13, 2014.
57. Mike Masnick, "Larry Lessig Launches Crowdfunded SuperPAC to Try to End SuperPACs," techdirt.com, May 1, 2014.
58. Jeremy Parish, "How Star Citizen Became the Most Successful Crowd Funded Game of All Time," wdc.com, January 13, 2015.
59. Rory Carroll, "Silicon Valley's Culture of Failure . . . and 'the Walking Dead' It Leaves Behind," theguardian.com, June 28, 2014.
60. Steven Poole, "What Does the Oculus Rift Backlash Tell Us? Facebook Just Isn't Cool," theguardian.com, March 27, 2014.
61. Nicholas Carson, "The Good, Bad, and Ugly of AngelList Syndicates," inc.com, September 30, 2013.
62. MicroVentures, "The 5 Keys to Becoming a MicroVentures Angel Investor" (sponsored content), venturebeat.com, January 1, 2014.
63. "How It Works," crowdfund.co/how-it-works/.
64. OneWorld South Asia, "Hand in Hand Creates 1.3 million Jobs," southasia.oneworld.net, February 26, 2013.

65. Matthew Ruben, "The Promise of Microfinance for Poverty Relief in the Developing World," seepnetwork.org, May 2007.
66. Amy Cortese, "Loans That Avoid Banks? Maybe Not," nytimes.com, May 3, 2014.
67. Amy Cortese, "Seeking Capital, Some Companies Turn to 'Do-It-Yourself I.P.O.'s,'" nytimes.com, July 31, 2013.
68. "Direct Public Offering," www.cuttingedgecapital.com/resources-and-links/direct-public-offering/.
69. Dan Rosenberg, "Raising Community Capital: Business Workshop on Direct Public Offerings," CommonBound Conference, Boston, June 6, 2014.
70. Ibid.
71. Babylonian Talmud, *Bava Metzia*, 42a.
72. Lietaer and Dunne, *Rethinking Money*.
73. Warren Buffett, quoted in Schuyler Velasco, "Warren Buffett: 10 Pieces of Investment Advice from America's Greatest Investor," *Christian Science Monitor*, August 30, 2013.
74. Tessa Hebb and Larry Beeferman, "U.S. Pension Funds' Labour-Friendly Investments," Alfred P. Sloan Industry Studies 2008, Annual Conference, May 1–2, 2008.
75. Ibid.
76. Hilaire Belloc, *The Servile State* (CreateSpace Independent Publishing Platform, 2008).
77. David Bolier, *Think Like a Commoner: A Short Introduction to the Life of the Commons* (Vancouver: New Society Publishers, 2014), 23.
78. Elinor Ostrom, *Governing the Commons: The Evolution of Institutions for Collective Action* (Cambridge: Cambridge University Press, 1990).
79. "Vandana Shiva on the Problem with Genetically Modified Seeds," *Moyers and Co.*, PBS, July 13, 2012.
80. opensourceecology.org/about-overview/.
81. "Machines: Global Village Construction Set," opensourceecology.org/gvcs/.
82. opensourceecology.org/about-overview/.
83. David Bollier, "The FLOK Society Vision of a Post-Capitalist Economy," bollier.org, March 2, 2014.
84. Bethany Horne, "How the FLOK Society Brings a Commons Approach to Ecuador's Economy," shareable.net, October 22, 2013.
85. Glyn Moody, "The FLOK Society Project: Making the Good Life Possible Through Good Knowledge," techdirt.com, June 11, 2014.
86. John Restakis, "The Emilian Model—Profile of a Co-operative Economy," British Columbia Co-operative Association, bcca.coop, 2000.
87. John Médaille, *Toward a Truly Free Market* (Wilmington, Del.: Intercollegiate Studies Institute Books, 2011).
88. Ibid.
89. Laura Flanders, "Talking with Chomsky," counterpunch.org, April 30, 2012.

90. Marie-Claire Kidd, "Plans for Co-Operative Alternative to eBay Take Shape," the news.coop, December 3, 2014.
91. "FAQ," www.fairmondo.de/faq#a1.
92. fairmondo.de.
93. Fairnopoly Team, "Dein Anteil," info.fairmondo.de, May 14, 2012.
94. "Projekt10000—Gemeinsam unsere Wirtschaft verändern," mitmachen.fair mondo.de/projekt10000/.
95. Felix Weth, "Genossenschaft 2.0," info.fairmondo.de, February 12, 2013.
96. Vasilis Kostakis and Michel Bauwens, *Network Society and Future Scenarios for a Collaborative Economy* (London: Palgrave Macmillan, 2014). A draft is available at p2pfoundation.net, December 30, 2014.
97. Amanda B. Johnson, "La'Zooz: The Decentralized Proof-of-Movement 'Uber' Unveiled," cointelegraph.com, October 19, 2014.
98. Trebor Scholz, "Platform Cooperativism vs. the Sharing Economy," medium.com, December 5, 2014.

Chapter Five: Distributed

1. Pope Francis, *Laudato Si'*, papal encyclical, Rome, 2015, paragraph 194.
2. Peter Beinart, "Jeb Bush Tries to Push the Pope Out of Politics," *The Atlantic*, June 17, 2015.
3. Pope Francis, *Laudato Si'*, paragraph 129.
4. Pope Leo XIII, *Rerum Novarum*, papal encyclical, Rome, 1891.
5. Pope Pius X, *Quadragesimo Anno*, papal encyclical, Rome, 1931.
6. Belloc, *Servile State*.
7. G. K. Chesterton, *Three Works on Distributism* (CreateSpace Independent Publishing Platform, 2009).
8. Janine Marchessault, *Marshall McLuhan* (London: Sage Publications, 2005).
9. Pope Leo XIII, *Rerum Novarum*.
10. Pope Francis, *Laudato Si'*, paragraph 114.

Selected Bibliography

Articles

Barber, Brad M. and Terrance Odean. "The Internet and the Investor." *Journal of Economic Perspectives* 15, no. 1 (Winter 2001).

Carlos, Ann M., and Stephen Nicholas. "'Giants of an Earlier Capitalism': The Chartered Trading Companies as Modern Multinationals." *Business History Review* 62, no. 3 (1988): 398–419.

Citizen Media Law Project. "Primer on Low-Profit Limited Liability Companies (L3Cs)." Berkman Center for Internet & Society at Harvard University. October 2010.

Even, William E., and David A. Macpherson. "Why Did Male Pension Coverage Decline in the 1980s?" *Industrial and Labor Relations Review* 47, no. 3 (April 1993).

Fantacci, Luca. "The Dual Currency System of Renaissance Europe." *Financial History Review* 15, no. 1 (2008).

Gatch, Loren. "Local Money in the United States During the Great Depression." *Essays in Economics & Business History* 26 (2008).

Gilbert, Clark, Matthew Eyring, and Richard N. Foster. "Two Routes to Resilience." *Harvard Business Review,* December 2012.

Hagel, John, III, et al. "Measuring the Forces of Long-term Change: The 2009 Shift Index." Deloitte, 2009.

Hauben, Michael, and Ronda Hauben. "Netizen: On the History and Impact of Usenet and the Internet." *First Monday: Peer-Reviewed Journal on the Internet* 3, no. 7 (July 6, 1998).

Huston, Larry, and Nabil Sakkab. "Connect and Develop: Inside Procter & Gamble's New Model for Innovation." *Harvard Business Review,* March 2006.

Page, Will, and Eric Garland. "The Long Tail of P2P." *Economic Insight,* no. 14 (May 9, 2014).

Rotman, David. "How Technology Is Destroying Jobs." *MIT Technology Review,* June 12, 2013.

Wallace, Bob. "AT&T Service Helps Broker Shave Costs." *Network World* 7, no. 31 (July 30, 1990): 13.

Protecting Pensions: Policy Analysis and Examples from OECD Countries, no. 8, Paris: OECD Publications, 2007.

Books

Antonopoulos, Andreas M. *Mastering Bitcoin: Unlocking Digital Cryptocurrencies.* Sebastopol, Calif. O'Reilly Media, 2014.

Belloc, Hillaire. *The Servile State.* CreateSpace Independent Publishing Platform, 2008.

Black, Edwin. *IBM and the Holocaust.* New York: Crown Publishers, 2001.

Bolier, David. *Think Like a Commoner: A Short Introduction to the Life of the Commons.* Vancouver: New Society Publishers, 2014.

Braudel, Fernand. *Civilization and Capitalism, 15th–18th Century,* vol. 1: *The Structure of Everyday Life.* New York: Harper & Row, 1982.

Brynjolfsson, Erik, and Andrew McAfee. *The Second Machine Age: Work, Progress, and Prosperity in a Time of Brilliant Technologies.* New York, London: W. W. Norton, 2014.

Chesterton, G. K. *Three Works on Distributism.* CreateSpace Independent Publishing Platform, 2009.

Cipolla, Carlo M. *Before the Industrial Revolution: European Society and Economy, 1000–1700.* 3rd ed. New York: W. W. Norton, 1994.

Daly, Herman. *Beyond Growth: The Economics of Sustainable Development.* Boston: Beacon Press, 1996.

Elberse, Anita. *Blockbusters: Hit-Making, Risk-Taking, and the Big Business of Entertainment.* New York: Henry Holt and Co., 2013.

Fisher, Irving. *The Purchasing Power of Money.* New York: Macmillan, 1920.

Franco, Pedro. *Understanding Bitcoin: Cryptography, Engineering and Economics.* New York: John Wiley & Sons, 2014.

Greco, Thomas H. *Understanding and Creating Alternatives to Legal Tender.* White River Junction, Vt.: Chelsea Green Publishing Co., 2001.

Hagel, John, John Seely Brown, and Duleesha Kulasooriya. *Shift Happens: How the World Is Changing, and What You Need to Do About It.* Houston, Tex.: Idea Bite Press, 2014.

Hartmann, Thom. *Unequal Protection: The Rise of Corporate Dominance and the Theft of Human Rights.* New York: Rodale Books, 2002.

Hebb, Tessa, and Larry Beeferman. "U.S. Pension Funds' Labour-Friendly Investments." Alfred P. Sloan Industry Studies 2008. Annual Conference, May 1–2, 2008.

Hoffman, Reid. *The Alliance: Managing Talent in a Networked Age*. Cambridge, Mass.: Harvard Business Review Press, 2014.

Keen, Andrew. *The Internet Is Not the Answer*. New York: Atlantic Monthly Press, 2015.

Kepos, Paula. *International Directory of Company Histories*, vol. 7. Farmington Hills, Mich.: St. James Press, 1993.

Klein, Naomi. *This Changes Everything: Capitalism vs. the Climate*. New York: Simon and Schuster, 2014.

Kostakis, Vasilis, and Michel Bauwens. *Network Society and Future Scenarios for a Collaborative Economy*. London: Palgrave Macmillan, 2014.

Lanier, Jaron. *Who Owns the Future?* New York: Simon and Schuster, 2013.

Lewis, Michael. *The Big Short: Inside the Doomsday Machine*. New York, London: W. W. Norton, 2011.

Lietaer, Bernard A., and Stephen M. Belgin. *Of Human Wealth: Beyond Greed & Scarcity*. Boulder, Colo: Human Wealth Books and Talks, 2001.

Lietaer, Bernard, and Jacqui Dunne. *Rethinking Money: How New Currencies Turn Scarcity into Prosperity*. San Francisco: Berrett-Koehler Publishers, 2013.

McLuhan, Marshall, and Eric McLuhan. *Laws of Media: The New Science*. Toronto: University of Toronto Press, 1992.

Marchessault, Janine. *Marshall McLuhan*. London: Sage Publications, 2005.

Médaille, John. *Toward a Truly Free Market*. Wilmington, Del.: Intercollegiate Studies Institute Books, 2011.

Mill, John Stuart. *Principles of Political Economy with Some of Their Applications to Social Philosophy*. London: Longmans, Green and Co., 1909.

Olen, Helaine. *Pound Foolish: Exposing the Dark Side of the Personal Finance Industry*. New York: Penguin/Portfolio, 2012.

Ostrom, Elinor. *Governing the Commons: The Evolution of Institutions for Collective Action*. Cambridge: Cambridge University Press, 1990.

Perez, Carlota. *Technological Revolutions and Financial Capital*. Cheltenham, England: Edward Elgar Press, 2002.

Piketty, Thomas. *Capital in the Twenty-First Century*. Translated by Arthur Goldhammer. Cambridge, Mass.: Belknap Press, 2014.

Raymond, Eric S. *The Cathedral and the Bazaar*. Sebastopol, Calif.: O'Reilly Media, 1999.

Rifkin, Jeremy. *The Zero Marginal Cost Society*. Basingstoke, England: Palgrave Macmillan, 2014.

Rushkoff, Douglas. *Life Inc.: How Corporations Conquered the World, and How We Can Take It Back*. New York: Random House, 2009.

———. *Present Shock: When Everything Happens Now*. New York: Penguin/Current, 2013.

Sacks, Justin. *The Money Trail: Measuring Your Impact on the Local Economy Using LM3*. London: New Economics Foundation & the Countryside Agency, December 2002.

Schumpeter, Joseph A. *Capitalism, Socialism and Democracy*. 3rd ed. New York: Harper Perennial, 2008.

Schor, Juliet B. *The Overworked American: The Unexpected Decline of Leisure*. New York: Basic Books, 1992.

———. *Plenitude: The New Economics of True Wealth*. New York: Penguin Press, 2010.

Shuman, Michael. *Local Dollars, Local Sense*. White River Junction, Vt.: Chelsea Green Publishing Co., 2012.

Simpson, Christopher. *Science of Coercion*. Oxford: Oxford University Press, 1996.

Tapscott, Don, and Anthony D. Williams. *Wikinomics: How Mass Collaboration Changes Everything*. New York: Penguin/Portfolio Trade, 2010.

Television

"Generation Like." *Frontline*. PBS. February 18, 2014.

Web Sites

Horne, Bethany. "How the FLOK Society Brings a Commons Approach to Ecuador's Economy." shareable.net. October 22, 2013.

Scholz, Trebor. "Platform Cooperativism vs. the Sharing Economy." medium.com. December 5, 2014.

U.S. Department of the Treasury. "History of 'In God We Trust.'" treasury.gov. March 8, 2011.

White Paper

Nakamoto, Satoshi. "Bitcoin: A Peer-to-Peer Electronic Cash System." bitcoin.org. October 31, 2008.

Index